Churchill Babington

An introductory lecture on archaeology

Churchill Babington

An introductory lecture on archaeology

ISBN/EAN: 9783741163685

Manufactured in Europe, USA, Canada, Australia, Japa

Cover: Foto ©berggeist007 / pixelio.de

Manufactured and distributed by brebook publishing software
(www.brebook.com)

Churchill Babington

An introductory lecture on archaeology

AN

INTRODUCTORY LECTURE

ON

ARCHÆOLOGY

𝕯𝖊𝖑𝖎𝖛𝖊𝖗𝖊𝖉 𝖇𝖊𝖋𝖔𝖗𝖊 𝖙𝖍𝖊 𝖀𝖓𝖎𝖛𝖊𝖗𝖘𝖎𝖙𝖞 𝖔𝖋 𝕮𝖆𝖒𝖇𝖗𝖎𝖉𝖌𝖊.

BY

CHURCHILL BABINGTON, B.D., F.L.S.

DISNEY PROFESSOR OF ARCHÆOLOGY, SENIOR FELLOW OF ST JOHN'S COLLEGE,
MEMBER OF THE ROYAL SOCIETY OF LITERATURE, OF THE NUMISMATIC
AND SYRO-EGYPTIAN SOCIETIES, HONORARY MEMBER OF THE
HISTORICO-THEOLOGICAL SOCIETY OF LEIPSIC, AND OF
THE ARCHÆOLOGICAL INSTITUTE OF ROME.

CAMBRIDGE:
DEIGHTON, BELL, AND CO.
LONDON: BELL AND DALDY.
1865.

PREFACE.

THE following Lecture was divided in the delivery
into two parts; illustrative specimens being exhibited after the conclusion of the delivery of
each portion. It has been suggested that I
should add in the form of notes a few books which
may prove useful to the students of particular
branches of Archæology; my best thanks are due
to the Rev. T. G. Bonney and the Rev. W. G.
Searle for their kind and valuable assistance in
drawing up certain of the lists. For ancient art
and archæology K. O. Müller's Manual, so often
referred to, will in general sufficiently indicate the
bibliography, and it is only in a few departments,
in numismatics more especially, that it has been
deemed necessary to add anything to his references. M. Labarte's Handbook, from which a great
part of the concluding portion of this lecture is
derived, will do the same thing, though in a far
less complete manner, for medieval art.

CONTENTS.

INTRODUCTORY LECTURE

ON

ARCHÆOLOGY.

FOLLOWING the example of my distinguished predecessor in the Disney Professorship of Archæology, I open my first Course of Lectures with an introductory Lecture on Archæology itself, so far as the very limited time for preparation has allowed me to attempt one.

I cannot indeed conceal from myself, and still less can I conceal from you, that no introductory Lecture which I could give, even if I were to take my own time in writing it, would bear any comparison with the compositions of his elegant and learned pen. It certainly does not proceed from flattery, and I hope not from an undue partiality of friendship to say of him, that in his power of grasping a complicated subject, of presenting it in a clear light, of illustrating it with varied learning, and of expressing himself in relation thereto in appropriate language, I have rarely seen his equal. To how great a disadvantage

1

then must I necessarily appear, when I have had
only six weeks' time in which to get ready this
as well as five other Lectures, and have been
moreover compelled to devote a considerable part
even of that short time to other and not less
important duties. A great unwillingness however that the Academical year should pass over
without any Archæological Lectures being delivered by the Disney Professor, has induced me to
make the attempt more quickly than would under
other circumstances have been desirable or even
justifiable; and I venture to hope that when allow-
ance is made for the exigency of the case, I shall
find in you, who have honoured this Lecture by
your presence, a clement and even an indulgent
audience.   .

In an introductory Lecture which deals with
generalities, it is hardly to be expected that I either
can say or ought to try to say much which is
absolutely new to any of my hearers; and I shall
not affect to say anything peculiarly striking, but
shall rather attempt to bring before you in a plain
way a view of the subject, which aims at being
concise and comprehensive; and in connexion
therewith respectfully to submit a few observa-
tions which have relation to other Academical
studies, as well as to the character of this particular Professorship.

What I propose then to do is this, first to explain what Archæology is; next to put in a clear
light what the character of this Professorship is;

after that to attempt a general sketch of the existing remains of Antiquity; then to point out the qualifications necessary or desirable for an archæologist; and in conclusion, to indicate the pleasure and advantage which flow from his pursuits.

The field of Archæology is vast, and almost boundless; the eye, even the most experienced eye, can hardly take in the whole prospect; and those who have most assiduously laboured in its exploration will be most ready to admit, that there are portions, and those large portions, which are to them either almost or altogether unknown.

For what is Archæology? It is, I conceive, the science of teaching history by its monuments[1], of whatever character those monuments may be. When I say history, I use the word not in the limited sense of the history of dynasties or of governments. Archæology does indeed concern itself with these, and splendidly does it illustrate and illuminate them; but it also concerns itself with every kind of monument of man which the ravages of time have spared.

---

[1] Perhaps it would be more correct to say 'by its *contemporary sensible* monuments,' so as to exclude later copies of ancient writings, or the *monumenta litterarum*, which fall more especially to the province of the scholar. A MS. of Aristotle of the thirteenth century is an archæological monument of that century only; it is a literary monument of the fourth century B.C. But a Greek epigram or epitaph which occurs on a sepulchral monument of the same or any other century B.C. is an archæological as well as a literary monument of that century.

Archæology concerns itself with the domestic
and the social, as well as with the religious, the
commercial, and the political life of all nations and
of all tribes in the ages that have passed away.
All that men in ancient times have made, and
left behind them, is the farrago of our study.

The archæologist will consequently make ob-
servations and speculations on the sites of ancient
cities where men have dwelt; on their walls and
buildings, sacred and profane; on their altars and
their market-places; on their subterranean constructions, whether sepulchres, treasuries, or drains.
He will trace the roads and the fosses along which
men of the old world moved, and on which
men often still move; he will explore the routes
of armies and the camps where they have pitched,
and will prowl about the barrows in which they
sleep;

    Excæ inveniet scabra robigine pila,
    Grandiaque effossis mirabitur ossa sepulchria.

He will also collect and classify every kind of
object, which man has made for use or for ornament in his own home, or in the city; in the fields,
or on the water. He will arrange the weapons
of offence and defence according to their material
and age; whether of stone, of bronze, of iron, or of
steel; among which some are so rude that a practised eye alone distinguishes them from the broken
flint stones lying in the field, others again so
elaborate as to rank among the most beautiful
productions both of classical and medieval art; he

will not disdain to preserve the bricks and the
tiles, which have once formed parts of Asiatic
cities or of Roman farms; he will excavate the
villas of the ancients; unearth their mosaic pavements; clean their lamps and candelabra; he will
mend or restore their broken crockery, and glass;
he will even penetrate into the lady's chamber,
turn over her toilet, admire her brooches and
her bracelets, examine her mirrors and her pins;
and all this he will do in addition to studying the
nobler works of ancient art, such as engraved gems
and medallions; works chased, carved and em-
bossed in the precious metals and in ivory; frescoes
and vase-paintings; bronzes and statues. He will,
likewise, familiarise himself with the alphabets of
the ancient nations, and exercise his ingenuity in
deciphering their written records, both public and
private; whether these be contained in inscriptions on stones or metal plates, or in papyrus-
rolls, or parchment books; or be scratched on
walls or on statues; or be painted on vases; or, in
fine, surround the device of a coin.

I have now mentioned some of the principal
objects of archæology, which, as I have said,
embraces within its range all the monuments of
the history and life of man in times past. And
this it does, beginning with the remains of primeval man, which stretch far beyond the records of all
literary history, and descending along the stream
of time till it approaches, but does not quite
reach time actually present. No sharp line of

demarcation separates the past from the present;
you may say that classical archæology termi-
nates with the overthrow of the Western Empire;
you may conceive that medieval archæology ceases
with the reign of Henry the Seventh; but, be this
as it may, in a very few generations the objects
of use or of ornament to us will become the
objects of research to the archæologist; and, I
may add, may be the subjects of lectures to my
successors.

For the founder of this Professorship, whose
memory is never to be named without honour,
and the University which accepted it, together
with his valuable collection of ancient sculptures,
undoubtedly intended that any kind or class of
antiquities whatever might fitly form the theme
of the Professor's discourse. I say this, because
a misconception has undoubtedly prevailed on
this subject, from which even my learned predecessor himself was not free. "Every nation of
course," says he, "has its own peculiar archæology. Whether civilized or uncivilized, whether
of historic fame or of obscure barbarism, Judæa,
Assyria, and. Egypt; Greece and Rome; India,
China, and Mexico; Denmark, Germany, Britain, and the other nations of modern Europe,
all have their archæology. The field of in-
quiry," he continues, "is boundless, and in the
multitude of objects presenting themselves the
enquirer is bewildered. It has been wisely pro-
vided therefore by the founder of this Profes-

sorship, that we shall direct our attention more
immediately to one particular class of Antiquities, and that the noblest and most important
of them all, I mean the Antiquities of Greece
and Rome'." Very probably such may have been
Mr Disney's original intention; and if so, this
will easily explain and abundantly pardon the
error of my accomplished friend; but the actual
words of the declaration and agreement between
Mr Disney and the University, which is of
course the only document of binding force, are
as follows: "That it shall be the duty of the
Professor to deliver in the course of each academical year, at such days and hours as the Vice-
Chancellor shall appoint, six lectures at least on
the subject of Classical, Mediæval and other Antiquities, the Fine Arts and all matters and things
connected therewith." Whether he would have
acted wisely or not wisely in limiting the field
to classical archæology, he has in point of fact not
thus limited it. And, upon the whole, I must
confess, I am glad that he has imposed no
limitation. For while there are but few who
would deny that many of the very choicest
relics of ancient art and of ancient history are
to be sought for in the Greek and Roman saloons
and cabinets of the museums of Europe, yet it
must at the same time be admitted that there are
other branches of archæology, which are far too

' Marsden's *Introd. Lect.* p. 5. Cambr. 1852.

important to be neglected, and which have an
interest, and often a very high interest, of their
own.

Let it be confessed, that the archæology of
Greece has in many respects the pre-eminence over
every other.   "It is to Greece that the whole
civilized world looks up," says Canon Marsden,
"as its teacher in literature and in art; and it is
to her productions that we refer as the standard of
all that is beautiful, noble, and excellent.   Greece
excelled in all that she put her hand to.   Her
sons were poets and orators and historians; they
were architects and sculptors and painters.   The
scantiest gleanings of her soil are superior to that
which constitutes the pride and boast of others.
Scarcely a fragment is picked up from the majestic
ruin, which does not induce a train of thought
upon the marvellous grace and beauty which must
have characterized the whole !

Quale te dicat tamen
Antehao fuisse, tales cum sint relliquiæ."

These eloquent and fervid words proceed
from a passionate admirer of Hellenic art, and
a most successful cultivator of its archæology.
Nor do I dare to say that the praise is exagge-
rated.   But at the same time, viewed in other
aspects, the archæology of our own country has
even greater interest and importance for us.   What
man is there, in whose breast glows a spark of
patriotism, who does not view the monuments of
his country which are everywhere spread around

him, (in this place above most places,) which connect the present with the remote past, and with
many and diverse ages of that past by a thousand
reminiscences, with feelings deeper and nobler
than any exotic remains of antiquity, how charm-
ing soever, could either foment or engender? This
love of national antiquities, seated in a healthy
patriotic feeling, has place in the speech of an
apostle himself: "Men and brethren, let me freely
speak unto you of the patriarch David, that he is
both dead and buried; and his sepulchre is with
us unto this day." The same feeling prompted
Wordsworth thus to express himself in reference
to our ancient colleges and their former occu-
pants:

> I could not always lightly pass
> Through the same gateways, sleep where they had slept,
> Wake where they waked; I could not always print
> Ground where the grass had yielded to the steps
> Of generations of illustrious men,
> Unmoved......
>   Their several memories here
> Put on a lowly and a touching grace
> Of more distinct humanity.

And not only the buildings, but the other
archæological monuments of the University (for
so I think I may be permitted to call the pictures
and the busts, and the statues, and the tombs,
which are the glories of our chapels, our libraries
and our halls) teach the same great lessons. They
raise up again our own worthies before our very
eyes, calling on us to strive to walk as they walk-

ed, dead though they be and buried; for their
effigies and their sepulchres are 'with us to this
day.' I must repeat, then, that I am glad that
the Disney Professor is not obliged to confine
himself to classical archæology, sorry as I should
be if he were wholly unable to give lectures on
one or more branches of that most interesting
department, which has moreover a special connexion with the classical studies of the University. It is manifest that the University intended
the Professor to consider no kind of human antiquities as alien from him; and I think this in
itself a very great gain. For, if the truth must
be confessed, antiquaries above most others have
been guilty of the error of despising those branches
of study which are not precisely their own. I
forbear to adduce proofs of this, though I am
not unprovided with them; and even although
you would certainly be amused if I were to read
them; classicists against gothicists; gothicists
against classicists.

I could wish that the learned and meritorious
writers on both sides had profited by the judicious
remarks of Mr Willson, prefixed to Mr Pugin's
Specimens of Gothic Architecture in England.
"The respective beauties and conveniences proper to the Grecian orders in their pure state or as
modified by the Romans and their successors in
the Palladian school may be fully allowed, without
a bigoted exclusion of the style we are accustomed
to term Gothic. Nor ought its merits to be as-

serted to the disadvantage of the classic style.
Each has its beauties, each has its proportions[1]."
One of the most eminent Gothic architects, Mr
George Gilbert Scott, expresses himself in a very
similar spirit. "It may be asked, what influence
do we expect that the present so called classic
styles will exercise upon the result we are imagin-
ing, (i. e. the developement of the architecture of
the future). Is the work of three centuries to be
unfelt in the future developements, and are its
monuments to remain among us in a state of
isolation, exercising no influence upon future art?
It would, I am convinced, be as unphilosophical
to wish, as it would be unreasonable to expect
this[2]." To turn from them to the classicists.
"See how much Athens gains," says Prof. T. L.
Donaldson, "upon the affections of every people, of
every age, by her Architectural ruins. Not a tra-
veller visits Greece whose chief purpose is not
centred in the Acropolis of Minerva .... But in
thus rendering the homage due to ancient Art it
were unjust to pass without notice those sublime
edifices due to the Genius of our Fathers. It is
now unnecessary to enter upon the question, whether the first ideas of Gothic Architecture were the
result of a casual combination of lines or a feli-
citous adaptation of form derived immediately

[1] P. xix. London, 1821.
[2] Scott's *Remarks on Secular and Domestic Architecture,
present and future*, p. 272. London, 1857.

from Nature: But graceful proportion, solemnity
of effect, variety of plan, playfulness of outline and
the profoundest elements of knowledge of construc-
tion place these edifices on a par with any of
ancient times. Less pure in conception and detail,
they excel in extent of plan and of disposition, and
yield not in the mysterious effect produced on the
feelings of the worshipper. The sculptured presence of the frowning Jove or the chryselephan-
tine statue of Minerva were necessary to awe
the Heathen into devotion. But the presence
of the Godhead appears, not materially but
spiritually, to pervade the whole atmosphere of
one of our Gothic Cathedrals[1]." The Editor of
The Museum of Classical Antiquities, well says,
"As antiquity embraces all knowledge, so in-
vestigations into it must be distinct and various.
Each antiquary labours for his own particular object, and each severally assists the other[2]." It
should be borne in mind moreover that archæological remains of every kind and sort are really
a part of human history; and if all parts of history
deserve to be studied, as they most assuredly do,
being parts, though not equally important parts,
of the Epic unity of our race, it will follow even
with mathematical precision that all monuments

[1] *Preliminary Discourse* pronounced before the University
College of London, upon the commencement of a series of Lectures on Architecture, pp. 17—24. London, 1842.

[2] *Museum of Classical Antiquities*, Vol. I. p. 1. London,
1851.

relating to all parts of that history must be worthy
of study also.

I desire therefore to express in language as
strong as may be consistent with propriety, my
entire disapproval of pitting one branch of archæology against another, or indeed any study against
another study. And on this very account I rejoice
that the Disney Professor's field of choice is as
wide as the world itself, so far as concerns its archæo-
ology. There is no country, there is no period
about which he may not occupy himself, or on which
he may not lecture, if he feel himself qualified to
do so. He is in a manner bound by the tenure
of his office to treat every branch of archæology
with honourable respect; and this in itself may
not be without a wholesome influence both upon
his words and sentiments. I have been somewhat
longer over this matter than I could have wished;
but I thought it desirable that the position of
the Disney Professor should be rightly under-
stood; and I have also endeavoured to shew the
real advantage of that position.

His field then is the world itself; but as this
is so (and as I think rightly so) there is a very true
and real danger lest he and his hearers should be
mazed and bewildered at the contemplation of its
magnitude. Yet in spite of that danger I will
venture to invite you to follow the outlines of the
great entirety of the relics of the ages that have
for ever passed away. I say the outlines, and
even this is almost too much, for I am compelled

to shade some parts of the picture so obscurely,
and to throw so much of other parts into the
background, that even of the outlines I can distinctly present to you but a portion. Thus I will
say little more of the archæology of the New
World, than that there is one which reaches far
beyond the period of Spanish conquest, comprising
among many other things ruins of Mexican cities,
exquisite monuments of bas-reliefs and other carvings in stone; I will not invite you into the far
East of the Old World, to explore the long walls
and Buddhist temples of the ancient and stationary
civilisation of China, or to dwell upon the objects
of its fictile and other arts; but leaving both
this and all the adjacent countries of Thibet,
Japan and even India without further notice, or
with only passing allusions, *spatiis conclusus ini-
quis*, I will endeavour, so far as my very limited
knowledge permits, the delineation of the most
salient peculiarities of the various remains of
the old world till the fall of the Roman Empire
in the West, and then attempt to trace briefly the
remains of successive medieval classes of antiquities, until we arrive at almost modern times.
I can name but few objects under each division of
the vast subject; but these will be selected so
as to suggest as much as possible others of a
kindred kind. In addressing myself to such an
audience, I may, if anywhere, act upon the as-
sumption, *Verbum sapienti sat est:* a single word
may suggest a train of thought. If I cannot

wholly escape the charge of tediousness, I must
still be content : for I am firmly convinced after the
most careful consideration that I can pursue no
course which is equally profitable, though I might
take many others which might be more amusing.

It would now appear probable that the earliest
extant remains of human handicraft or skill have
as yet been found, not on the banks of the Nile
or the Euphrates, but in the drift and in the
caverns of Western Europe. Only yesterday, as
I may say, it has been found out that in a geological period when the reindeer was the denizen
of Southern France, and when the climate was
possibly arctic, there dwelt in the caverns of the
Périgord a race of men, who were unacquainted
with the use of metals, but who made flint and
bone weapons and instruments ; who lived by
fishing and the chase, eating the flesh of the
reindeer, the aurochs, the wild goat and the
chamois ; using their skins for clothes which they
stitched with bone needles, and their bones for
weapon handles, on which they have etched representations of the animals themselves. Specimens
of these things were placed last year in the
British Museum ; and a full account of the dis-
coveries in 1862 and 1863 may be seen in the
Revue Archéologique. Some distinguished antiquaries consider that they are the earliest human
remains in Western Europe. Various other dis-
coveries in the same regions of late years have
tended towards shewing that the time during

which man has lived upon the earth is much
greater than we had commonly supposed. The
geological and archæological circumstances under
which the flint implements were found at Abbe-
ville, and St Acheul, near Amiens, in the valley of
the Somme, left no doubt that they were anterior
by many ages to the Roman Empire. They have
a few points of similarity to those found in the
caverns of the Périgord, and as they occur along
with the remains of the Elephas antiquus and the
hippopotamus, Sir Charles Lyell infers that both
these animals coexisted with man; and perhaps
on the whole we may consider them rather than
those of the Périgord to be the earliest European
remains of man, or of man at all. Similar weapons
have been found in the drift in this country, in
Suffolk, Bedfordshire, and elsewhere. At Brixham,
near Torquay, a cavern was examined in 1858,
covered with a floor of stalagmite, in which were
imbedded bones of the reindeer and also an entire
hind leg of the extinct cave-bear, every bone of
which was in its proper place; the leg must consequently have been deposited there when the separate bones were held together by their ligaments.
Below this floor was a mass of loam or bone-
earth, varying from one to fifteen feet in thickness,
and amongst it, and the gravel lying below it,
were discovered about fifteen flint knives, recog-
nised by practised archæologists as artificially
formed, and among them one very perfect tool
close to the leg of the bear. It thus becomes

manifest that the extinct bear lived after the flint
tools were made, or at any rate not earlier; so
that man in this district was either the contemporary of the cave-bear, or (as would seem
more probable) his predecessor. But shortness
of time forbids me to do more than to indicate
that in western Europe generally, as well as in
Britain, we have an archæology beginning with
the age of the extinct animals or quaternary
geological epoch and connecting itself with the
age of the Roman Empire, when the first literary
notices of those countries, with slight exceptions,
commence. The antiquaries and naturalists of
Denmark conjointly (these indeed should always be
united, having much in common; and I am happy
in being able to say that a love of archæology
has often been united with a love of natural
science by members of this University, among
whom the late and the present Professor of
Botany may be quoted as examples)—these Danish
archæologists and naturalists I say, have made
out three distinct periods during this interval: the
age of stone contemporary with the pine forests;
the age of bronze commencing with the oak
forests which lie over the pine in the peat; and
the age of iron co-extensive with the beech forests
which succeeded the oak, and which covered the
country in the Roman times as they cover it now.
The skulls belonging to the oldest or stone age re-
semble those of the modern Laplanders; those of
the second and third are of a more elongated type.

2

The refuse-heaps along the shores of the islands of the Baltic, consisting of the remains of mollusks and vertebrated animals, mingled with stone weapons, prove the great antiquity of the age of stone; the oyster then flourished in places where, by reason of the exclusion of the ocean from the brackish Baltic, it does not now exist. None of the animals now extinct, however, occur in these Kjökkenmödding, as they are called, except the wild bull, the *Bos primigenius*, which was alive in Roman times; but the bones of the auk, now, in all probability, extinct in Europe, are frequent; also those of the capercailzie, now very rare in the southern districts of Scandinavia, though abundant in Norway, which would find abundant food in the buds of the pines growing in prehistoric times in the peat bogs. Similar refuse-heaps, left in Massachusetts and in Georgia by the North American Indians, are considered by Sir C. Lyell, who has seen them, to have been there for centuries before the white man arrived. They have also been found, I understand, very recently in Scotland in Caithness. The stone weapons have now been sharpened by rubbing, and are less rude and probably more recent than those of the drift of the Somme valley, or of the caverns of the Périgord. The only domestic animal belonging to the stone age, yet found in Scandinavia, is the dog; and even this appears to have been wanting in France. In the ages of bronze and iron various domestic animals existed;

but no cereal grains, as it would seem, in the whole of Scandinavia. Weapons and tools belonging to these three periods, as well as fragments of pottery and other articles, are very widely diffused over Europe, and have been met with in great abundance in our own country (in Ireland more especially), as well as near the Swiss-lake habitations, built on piles, to which attention has only been called since 1853. It is strange that all the Lake settlements of the bronze period are confined to West and Central Switzerland: in the more Eastern Lakes those of the stone period alone have been discovered.

Similar habitations of a Pœonian tribe dwelling in Lake Prasias, in modern Roumelia, are mentioned by Herodotus, and they may be compared, in some degree, with the Irish Lake-dwellings or Crannoges, *i.e.* artificial islands, and more especially with the stockaded islands, occurring in various parts of the country: and which are accompanied by the weapons and instruments and pottery of the three aforesaid periods. Even in England slight traces of similar dwellings have been found near Thetford, not accompanied by any antiquities, but by the bones of various animals, the goat, the pig, the red deer, and the extinct ox, the *Bos longifrons*, the skulls of which last were in almost all instances fractured by the butcher.

As to the chronology and duration of the three periods I shall say nothing, though not ignorant that some attempts have been made to determine

2—2

them. They must have comprehended several
thousand years, but how many seems at present
extremely uncertain. I should perhaps say that
Greek coins of Marseilles, which would probably
be of the age of the Roman Republic, have been
found in Switzerland in some few aquatic stations,
and in tumuli among bronze and iron implements
mixed. The cereals wanting in Scandinavia appear in Switzerland from the most remote period;
and domestic animals, the ox, sheep, and goat, as
well as the dog, even in the earliest stone-settle-
ments. Among the ancient mounds of the valley
of the Ohio, in North America, have been found
(besides pottery and sculpture and various articles
in silver and copper) stone weapons much resem-
bling those discovered in France and other places
in Europe. Before passing from these pre-historic
remains, as they are badly called, to the historic,
let me beg you to observe a striking illustration
of the relation of archæology to history. Archæology is not the handmaid of history; she occupies
a far higher position than that: archæology is, as I
said at the outset, the science of teaching history by
its monuments. Now for all western and northern
Europe nearly the whole of its early history must
be deduced, so far as it can be deduced at all,
from the monuments themselves; for the so-called
monuments of literature afford scanty aid, and for
that reason our knowledge of these early ages is
necessarily very incomplete. Doubtless, many a
brave Hector and many a brave Agamemnon lived,

fought, and died in the ages of stone and of
bronze; but they are oppressed in eternal night,
unwept and unknown, because no Scandinavian
Homer has recorded their illustrious deeds. Still,
we must be thankful for what we can get; and if
archæological remains (on which not a letter of an
alphabet is inscribed) cannot tell us everything,
yet, at least, everything that we do know about
these ages, or very nearly so, is deduced by archæology alone.

We must now take a few rapid glances at the
remains of the great civilised nations of the ancient
world. Mr Kenrick observes that the seats of its
earliest civilisation extend across southern Asia in
a chain, of which China forms the Eastern, and
Egypt the Western extremity; Syria, Mesopota-
mia, Assyria, and India, are the intermediate links.
In all these countries, when they become known
to us, we find the people cultivating the soil,
dwelling in cities, and practising the mechanical
arts, while their neighbours lie in barbarism and
ignorance. We cannot, he thinks, fix by direct
historical evidence the transmission of this earliest
civilisation from one country to another. But we
may determine with which of them ancient history
and archæology must begin. The monuments of
Egypt surpass those of all the rest, as it would
appear, by many centuries. None of the others
exercised much influence on European civilisation
till a later period, some exception being made for
the Phœnician commerce; but the connection of

European with Egyptian civilisation is both direct
and important. "From Egypt," he remarks,
"it came to Greece, from Greece to Rome, from
Rome to the remoter nations of the West, by
whom it has been carried throughout the globe[1]."
As regards its archæology, which is very peculiar and indeed in some respects unique, I must
now say a few words. The present remains of
Memphis, the earliest capital, said to have been
founded by Athothis, the son of Menes, the first
king of the first dynasty, are not great; but so
late as the fourteenth century they were very considerable. Temples and gateways, colossal statues
and colossal lions then existed, which are now
no more. Whether any of them approached the
date of the foundation it is useless to enquire.
Now, the most remarkable relic is a colossal statue
of Rameses II., which, when perfect, must have
been about forty-three feet high. This monarch is
of the XVIIIth dynasty, which embraces the most
splendid and flourishing period of Egyptian history; and though much uncertainty still prevails
for the early Egyptian chronology, it appears to
be well made out and agreed that this dynasty
began to reign about fifteen centuries before the
Christian era. But the pyramids and tombs of
Ghizeh, and of several other places at no great
distance from Memphis, are of a much earlier
date; and the great pyramid is securely referred

[1] *Ancient Egypt*, Vol. I. p. 3. London, 1850.

to a king of the fourth dynasty. "Probably at no place in the entire history of Egypt," says Mr Osburn, "do the lists and the Greek authors harmonize better with the historical notices on the monuments than at the commencement of this dynasty[1]." The system of hieroglyphic writing was the same (according to Mr Kenrick) in all its leading peculiarities, as it continued to the end of the monarchy. I regret to say that some eminent men have tried to throw discredit, and even ridicule, on the attempts which, I think, have been most laudably made with great patience, great acuteness, and great learning, to decipher and interpret the Egyptian and other ancient languages. Many of us, doubtless, have seen a piece of pleasantry in which *Heigh-diddle-diddle, The cat and the fiddle* is treated as an unknown language; the letters are divided into words—all wrongly, of course—these words are analysed with a great show of erudition, and a literal Latin version accompanies the whole. If I remember (for I have mislaid the amusing production) it proves to be an invocation of the gods, to be used at a sacrifice. Now, a joke is a good thing in its place; only do not let it be made too much of. Every archæologist, beginning with Jonathan Oldbuck, must sometimes fall into blunders, when he takes inscriptions in hand, even if the language be a known one; and, of course, *à fortiori*, when

---

[1] *Monumental History of Egypt*, Vol. i. p. 262. London, 1854.

but little known. My own opinion on hieroglyphics would be of no value whatever, as I know
nothing beyond what I have read in a few modern
authors, and have never studied the subject; but,
allow me to observe, that I had a conversation
very lately with my learned and excellent friend,
Dr Birch, of the British Museum, who is now
engaged in making a dictionary of hieroglyphics,
and he assured me that a real progress has been
made in the study of them, that a great deal of
certainty has been attained to; while there is still
much that requires further elucidation. To the
judgment of such a man, who has spent a great
part of his life in the study of Egyptian antiquities, though he has splendidly illustrated other
antiquities also, I must think that greater weight
should be attached than to the judgment of others,
eminent as they may be in some branches of learning, who have never studied this as a specialty.

The relation of archæology to Egyptian history
deserves especial notice. We have not here, as in
pre-historic Europe, a mere multitude of uninscribed and inconsiderable remains; but we have
colossal monuments of all kinds—temples, gateways, obelisks, statues, rock sculptures—more or
less over-written with hieroglyphics; also sepul-
chral-chambers, in many instances covered with
paintings, in addition to a variety of smaller
works, mummy cases, jewelry, scarabæi, pottery,
&c., upon many of which are inscriptions. By aid
of these monuments mostly, but by no means ex-

clusively, the history of the Pharaohs and the
manners and customs of their people are recovered. The *monumenta litterarum* themselves are frequently preserved on the monuments of stone and
other materials.

For the pyramids of Ghizeh and the adjoining
districts, for the glorious temples of Dendera, of
Karnak, the grandest of all the remains of the
Pharaohs, as well as for those of Luxor, with its
now one obelisk, of Thebes, of Edfou, of Philœ,
likewise for the grottoes of Benihassan, I must
leave you to your own imagination or recollection, which may be aided in some degree by
a few of the beautiful photographs by Bedford,
which are now before your eyes. They extend
along the banks and region of the Nile—for this
is Egypt—from the earliest times down to the age
of the Ptolemies and of Cleopatra herself, and
even of the Roman empire, in the case of Dendera, where the portico was added by Tiberius to
Cleopatra's temple. Before quitting these regions
I would remark, that the extraordinary rock-hewn
temple of Aboo-Simbel in Nubia, which includes
the most beautiful colossal statues yet found—their
height as they sit is more than fifty feet—bears
some similarity to certain Indian temples, especially to the temple of Siva at Tinnevelly, and the
Kylas at Ellora, which last has excited the asto-
nishment of all travellers. "Undoubtedly," says
Mr Fergusson, "there are many very striking
points of resemblance...but, on the other hand,

the two styles differ so widely in details and in
purpose, that we cannot positively assert the
actual connexion between them, which at first
sight seems unquestionable[1]."

The archæology of the Babylonian empire need
only occupy a few moments. The antiquity of
Babylon is proved to be as remote as the fifteenth
century B. C., by the occurrence of the name on
a monument of Thothmes III., an Egyptian monarch of the XVIIIth dynasty. It may be much
older than that; but the archæological remains of
the Birs Nimroud (which was long imagined to
be the tower of Babel) hitherto found are not
older than the age of Nebuchadnezzar. This
palatial structure consisted, in Mr Layard's opinion,
of successive horizontal terraces, rising one above
another like steps in a staircase. Every inscribed
brick taken from it,—and there are thousands and
tens of thousands of these,—bears the name of
Nebuchadnezzar. It is indeed possible that he
may have added to an older structure, or rebuilt
it; and if so we may one day find more ancient
relics in the Birs. But at a place called Mujelibé
(the Overturned) are remains of a Babylonian
palace not covered by soil, also abounding with
Nebuchadnezzar's bricks, where Mr Layard found
one solitary fragment of a sculptured slab, having
representations of gods in head-dresses of the
Assyrian fashion, and indicating that the Baby-

[1] *Handbook of Architecture*, p. 101. London, 1859.

lonian palaces were probably similarly ornamented.
A very curious tablet was also brought from
Bagdad of the age of Nebuchadnezzar, giving,
according to Dr Hincks, an account of the temples which he built. Besides these, "a few inscribed tablets of stone and baked clay, figures in
bronze and terra cotta, metal objects of various
kinds, and many engraved cylinders and gems are
almost the only undoubted Babylonian antiquities
hitherto brought to Europe." Babylonia abounds
in remains, but they are so mixed—Babylonian,
Greek, Roman, Arsacian, Sassanian, and Christian
—that it is hard to separate them. Scarcely more
than one or two stone figures or slabs have been
dug out of the vast mass of débris; and, as Isaiah
has said, "Babylon is fallen, is fallen; and all the
graven images of her gods hath Jehovah broken
unto the ground[1]."

The most splendid archæological discovery of
our age is the disinterment of the various palaces
and other monuments of the Assyrian Empire.
The labours of Mr Layard and M. Botta have
made ancient Assyria rise before our eyes in all
its grandeur and in all its atrocity. In visiting
the British Museum we seem to live again in
ancient Nineveh. We behold the sculptured slabs
of its palaces, on which the history of the nation
is both represented and written; we wonder at its

[1] See Layard's *Nineveh and Babylon*, chapters xxii. xxiii.,
especially pp. 504, 528, 532. London, 1853.

strange compound divinities, its obelisks, its
elegant productions in metal, in ivory, and in
terra cotta. By patient and laborious attention to
the cuneiform inscriptions, aided by the notices
in ancient authors, sacred and profane, men like
Sir H. Rawlinson and Dr Hincks have recovered
something like a succession of Assyrian kings,
ranging from about 1250 B.C. to about 600 B.C.,
and many particulars of their reigns, some of
which bring out in a distinct manner the accurate
knowledge of the writers of the Old Testament.

The remains of ancient Persia are too considerable to be passed over. Among other monuments
at Pasargadæ, a city of the early Persians, is
a great monolith, on which is a bas-relief, and
a cuneiform inscription above, "I am Cyrus the
king, the Achæmenian." Here is the tomb of
the founder of the empire.

At Susa, the winter seat of the Persian kings
from the time of Cyrus, Mr Loftus and Sir W. F.
Williams have found noble marble structures raised
by Darius, the son of Hystaspes (524—485 B.C.),
whose great palace was here: commenced by himself
and completed by Artaxerxes II. or Mnemon (405—
359 B.C.). Both here and at Persepolis, the richest
city after Susa (destroyed, as we all remember
from Dryden's ode, by Alexander), are ruins of
magnificent columns of the most elaborate orna-
mentation, and many cuneiform inscriptions, deci-
phered by Lassen and Rawlinson. Mr Loftus
remarks on the great similarity of the buildings

of Persepolis and Susa, which form a distinct
style of architecture. This is the salient feature
of Persian archæology, and to him I refer you
upon it¹. I cannot dwell upon other ruins in these
regions, or on the minor objects, coins, cylinders,
and vases of the ancient Persian empire; and still
less on the very numerous coins of the Arsacidæ,
and Sassanidæ, who afterwards succeeded to it.

Of ancient Judæa we possess as yet very
scanty archæological monuments indeed before the
fall of the monarchy. The so-called Tombs of
the Kings are now, I believe, generally considered
to belong to the Herodian period. Of the Temple
of Jerusalem, the holy place of the Tabernacle of
the Most Highest, not one stone is left upon
another. And we may well conceive that nothing
less than its destruction would effectually convince
the world of the great truth that an hour had
arrived in which neither that holy mountain on
which it was built, nor any other in the whole
world, was to be the scene of the exclusive worship
of the Father. The sites of the Holy Places,
however, have naturally excited much attention,
and have been well illustrated by several distinguished resident members of our University, and
also by a foreign gentleman who for some time
resided among us. Dr Pierotti had the singular

¹ See his *Travels and Researches in Chaldæa and Susiana*,
ch. xxviii. London, 1857; also Smith's *Dict. of Greek and
Roman Geography*, s. v. Pasargadæ, Persepolis, Susa; and
Vaux's *Nineveh and Persepolis.* London, 1850.

good fortune to discover the subterranean drains
by which the blood of the victims, slaughtered in
the Temple, was carried off; and this discovery
afforded valuable aid in determining various previously disputed matters in connexion with the
Temple. He likewise came upon some masonry in
the form of bevelled stones below the surface,
which was not unreasonably supposed to belong to
Solomon's Temple; but it now appears that this
opinion is doubtful. Besides these, we have the
sepulchres of the patriarchs at Hebron, guarded
with scrupulous jealousy; and tanks at the same
place, which may be as old as the time of David,
and perhaps one or two things more of a similar
kind. We may well hope that the explorations
which are now being set on foot for bringing to
light the antiquities of Palestine may add to their
number.

In the relation of Jewish archæology to Jewish
history we have a case quite different to all those
that have gone before it: there the native archæology was more or less extensive, the independent
native literature scanty or non-existent; here,
where the archæology is almost blotted out, is it
precisely the reverse. We have in the sacred
books of the Old Testament an ample literary
history: we have scarcely any monumental remains of regal Judæa at all. With regard to
the New Testament the matter is otherwise;
archæological illustrations, as well as literary,
exist in abundance, and some very striking proofs

from archæology have been adduced of the veracity and trustworthiness of its authors. My predecessor bestowed great attention on the numismatic and other monumental illustrations of Scripture, and herein set a good example to all that should come after him. Archæology is worthily employed in illustrating every kind of ancient literature; most worthily of all does she occupy herself in the illustration and explanation and confirmation of the sacred writings, of the Book of books.

The antiquities of Phœnicia need not detain us long. Opposite to Aradus is an open quadrangular enclosure, excavated in rock, with a throne in the centre for the worship of Astarte and Melkarth; this is the only Phœnician temple discovered in Phœnicia, except a small monolithal temple at Ornithopolis, about nine miles from Tyre, of high antiquity, dedicated apparently to Astarte. I wish however to direct your attention to the characteristic feature of Phœnician architecture, its enormous blocks of stone bevelled at the joints. You have them in the walls of Aradus and in other places in Phœnicia. They are also found in the temple of the Sun at Baalbec, and may with great probability, I conceive, be regarded as Phœnician; though the rest of the beautiful architectural remains there are Greco-Roman of the Imperial period, and perhaps the best specimens of their kind in existence. Among other Phœnician antiquities we have sarcophagi, and sepulchral

chambers for receiving them, also very beautiful
variegated glass found over a good part of Europe
and Asia, commonly called Greek, but perhaps
more reasonably presumed to be Phœnician. Most
of the remains found on the sites of the Phœnician
settlements are either so late Phœnician, or so little
Phœnician at all, as at Carthage, that I shall make
no apology for passing over both them, and the few
exceptions also, just alluding however to the existence of a remarkable hypæthral temple in Malta,
which I myself saw nearly twenty years ago, not
long, I believe, after it was uncovered. With regard
to the strange vaulted towers of Sardinia, called
Nuraggis, they may be Phœnician or Carthaginian,
but their origin is uncertain. "All Phœnician
monuments," says Mr Kenrick, "in countries un-
questionably occupied by the Phœnicians are recent[1]." He makes the remark in reference to the
Lycian archæology. Whether the Lycians were of
Phœnician origin or not, their rock-temples and
rock-tombs, abounding in sculptures (illustrative
both of their mythology and military history),
shew that they were not much behind the Greeks in
the arts. With the general appearance of their
Gothic-like architecture, and of their strange
bilingual inscriptions, Greek and Lycian, we are
of course familiarised by the Lycian Room in the
British Museum. With regard to the relation of
Phœnician and Lycian archæology to the history

[1] *Phœnicia*, p. 88. London, 1855. See also Smith's *Dict.
of Greek and Roman Geography*, s. v. Phœnicia and Lycia.

of the peoples themselves, it must be sufficient to say, that their history, both literary and monumental, is quite fragmentary; in the case of Phœnicia the literary notices perhaps preserve more to us than the monumental; in regard to Lycia the remark must rather be reversed.

From Phœnicia, which first carried letters to Greece, let us also pass to Greece. But Greece, in the sense in which I shall use it, includes not only Greece Proper, but many parts of Asia Minor, as well as Sicily and the Great Greece of Italy. And here I must unwillingly be brief, and make the splendid extract from Canon Marsden, quoted before, in some degree do duty for me. But think for a minute first on its architecture, I do not mean its earliest remains, such as the Cyclopian walls and the lion-gate at Mycenæ, and the so-called treasury of Atreus, which ascend to the heroic ages or farther back, but its temple architecture. Before I can name them, images of the Parthenon, the Erectheum, the temple of Jupiter Panhellenius at Ægina, the temple of Apollo Epicurius at Phigalia or Bassæ, that of Concord (so called) at Agrigentum, the most perfect in Sicily, the three glorious Doric temples of Pæstum, the Ionic ruins of Branchidœ, will, I am confident, have arisen before your eyes. Many of us perhaps have seen some of them; if not, we all feel as though we had. Think of its sepulchral monuments, which are in the form of temples; and first of Queen Artemisia's Mausoleum, the most

3

splendid architectural expression of conjugal affec-
tion that has ever existed, the wonder of the
world, with its colossal statue of her husband
and its bas-reliefs by Bryaxis and Scopas and other
principal sculptors; and remember that we have
these in our national museum.   Various fine rock-
tombs, likewise in the form of temples, occur in
Asia Minor, *e.g.* that of Midas at Nacoleia, the
Lion-tomb at Cnidus, the necropolis at Telmessus.

The transition from temples and tombs to statuary is easy, as these were more or less decorated
with its aid.   Although we still possess the great
compositions of some of the first sculptors and brass-
casters, for example, the Quoit-thrower of Myron,
the Diadumenos of Polycleitus, (*i. e.* a youth binding his head with a fillet in token of an athletic
victory,) and perhaps several of the Venuses of
Praxiteles; yet it is needless for me to remind you
that these with few exceptions are considered to be
copies, not originals.   But yet there are exceptions.
"The extant relics of Greek sculpture," says Mr
Bunbury, "few and fragmentary as they undoubtedly are, are yet in some degree sufficient to
enable us to judge of the works of the ancient
masters in this branch of art.   The metopes of Seli-
nus, the Æginetan, the Elgin, and the Phigaleian
marbles, to which we now add the noble fragments
recently brought to this country from Halicar-
nassus, not only serve to give us a clear and defi-
nite idea of the progress of the art of sculpture,
but enable us to estimate for ourselves the mighty

works which were so celebrated in antiquity[1]."
Of bronzes of the genuine Greek period, which
we may call their metal statuary, the most beautiful that occur to my remembrance are those of
Siris, now in the British Museum. They are considered by Brönsted to agree in the most remarkable and striking manner with the distinctive
character of the school of Lysippus. But most
of the extant bronzes are, I believe, of the Roman
period, executed however, like their other best
works, by Græco-Roman artists.

With the Greek schools of painting, Attic,
Asiatic, and Sicyonian, no less celebrated than
their sculpture, it has fared far worse. There is
not one of their works surviving; no, not one. Of
these schools and their paintings I need not here
say anything, as I am concerned only with the
archæological monuments which are now in existence. But the loss is compensated in some
degree by the paintings on vases, in which we
may one day recognise the compositions of the
various great masters of the different schools,
just as in the majolica and other wares of the

---

[1] *Edinburgh Review* for 1858, Vol. CVIII. p. 382. I follow
common fame in assigning this article to Mr Dunbury; few
others indeed were capable of writing it. Besides the sculptures named by him we have in the British Museum a bas-
relief by Scopas, as it is thought, who may also be the author
of the Niobid group at Florence; likewise the Cares (so-called)
from Eleusis, and the statue of Pan from Athens, now in our
Fitzwilliam Museum. For other antique statues and bronzes
and for the later copies see Müller's *Ancient Art*, passim.

16th and following centuries we have the compositions of Raffaelle, Giulio Romano, and other painters. "The glorious art of the Greek painters," says K. O. Müller, the greatest authority for ancient art generally, "as far as regards light, tone, and local colours, is wholly lost to us; and we know nothing of it except from obscure notices and later imitations ;" (referring, I suppose, to the frescoes of Herculaneum and of Pompeii more especially ;) "on the contrary, the pictures on vases with thinly scattered bright figures give us the most exalted idea of the progress and achievements of the art of design, if we venture, from the workmanship of common handicraftsmen, to draw conclusions as to the works of the first artists[1]." But of this matter and of the vases themselves, which rank among the most graceful remains of Greek antiquity, and are found over the whole Greek world, I shall say no more now, as they will form the subject of my following lectures. We have also many terra cottas of delicate Greek workmanship, mostly plain, but some gilded, others painted, from Athens, as well as from a great variety of other places, of which the finest are now at Munich. Relief ornaments, sometimes of great beauty, in the same material, were impressed with moulds, and Cicero, in a letter to

[1] *Ancient Art and its Remains*, p. 119. Translated (with additions from Welcker) by Leitch. London, 1852. This invaluable work is a perfect thesaurus for the student, and will conduct him to the most trustworthy authorities on every branch of the subject.

Atticus, wishes for such *typi* from Athens, in order
to fix them on the plaster of an atrium. Most of
those which now remain seem to be of Greco-
Roman times.

Of the art of coinage invented by the Greeks
and carried by them to the highest perfection
which it has ever attained, a few words must now
be said. The history of a nation, said the first Na-
poleon, is its coinage: and the art which the Greeks
invented became soon afterwards, and now is, the
history of the world. Numismatics are the epitome of all archæological knowledge, and any one
who is versed in this study must by necessity be
more or less acquainted with many others also.
Architecture, sculpture, iconography, topography,
palæography, the public and private life of the
ancients and their mythology, are all illustrated
by numismatics, and reciprocally illustrate them.

Numismatics give us also the succession of kings
and tyrants over the whole Greek world. In the
case of Bactria or Bactriana, whose capital Bactra
is the modern Balk, this value of numismatics is
perhaps most conspicuous. From coins, and from
coins almost alone, we obtain the succession of
kings, beginning with the Greek series in the
. third century B.C., and going on with various
dynasties of Indian language and religion, till we
come down to the Mohammedan conquest. "Ex-
tending through a period of more than fifteen centuries," says Professor H. H. Wilson, "they furnish
a distinct outline of the great political and reli-

gious vicissitudes of an important division of
India, respecting which written records are im-
perfect or deficient[1]."

Coins are so much more durable than most
other monuments, that they frequently survive,
when the rest have perished. This is well put
by Pope in his Epistle to Addison, on his Discourse on Medals:

> Ambition sighed, she saw it vain to trust
> The faithless column and the crumbling bust,
> Huge moles whose shadows stretched from shore to shore,
> Their ruins perished and their place no more.
> Convinced she now contracts her vast design,
> And all her triumphs shrink into a coin.
> A narrow orb each crowded conquest keeps,
> Beneath her palm here sad Judea weeps;
> Now scantier limits the proud arch confine;
> And scarce are seen the prostrate Nile or Rhine;
> A small Euphrates thro' the piece is rolled,
> And little eagles wave their wings in gold.
> The Medal, faithful to its charge of fame,
> Through climes and ages bears each form and name;
> In one short view subjected to our eye,
> Gods, emperors, heroes, sages, beauties, lie.

Regarded simply as works of art the coins of
Magna Græcia and Sicily, more especially those
of Syracuse and its tyrants, as well as those of
Thasos, Opus, and Elis, also the regal coins of
Philip, Alexander, Mithridates, and some of the
Seleucidæ, are amongst the most exquisite produc-

---

[1] *Ariana Antiqua*, p. 439. London, 1841. For the more
recent views of English and German numismatists on these
coins, see Mr Thomas's *Catalogue of Bactrian Coins* in the
Numismatic Chronicle for 1857, Vol. XIX. p. 13 sqq.

tions of antiquity. Not even in gem-engraving, an art derived by Greece from Egypt and Assyria, but carried by her to the highest conceivable perfection, do we find anything superior to these. I must, before quitting the subject of numismatics, congratulate the University on the acquisition of one of the largest and most carefully selected private collections of Greek coins ever formed, viz. the cabinet of the late Col. Leake, which is now one of the principal treasures of the Fitzwilliam Museum.

Inferior as gems are to coins in most archæological respects, especially in respect of their connection with literary history, and though not superior to the best of them artistically, gems have nevertheless one advantage over coins, that they are commonly quite uninjured by time. Occasionally (it is true) this is the case with coins; but with gems it is the rule. Of course, to speak generally, the art of gems, whose material is always more or less precious, is superior to that of coins, which were often carelessly executed, as being merely designed for a medium of commercial exchange. High art would not usually spend itself upon small copper money, but be reserved for the more valuable pieces, especially those of gold and silver[1]. The subjects of gems are mostly mythological, or are connected with the heroic cycle; a smaller, but more inter-

---

[1] This remark however must not be pressed too closely. Certain small Greek copper coins of Italy, Sicily, &c., are exceedingly beautiful.

esting number, presents us with portraits, which
however are in general uninscribed. At the same
time, by comparing these with portrait-statues and
coins we are able to identify Socrates, Plato, Aristotle, Demosthenes, Alexander the Great, several of the Ptolemies, and a few others; most of
which may have been engraved by Greco-Roman
artists. But the catalogue of authentic portraits
preserved to us, both Greek and Roman, is, as
K. O. Müller observes, now very much to be
thinned.

With regard to ancient iconography in general,
coins, without doubt, afford the greatest aid; but
no certain coin-portraits are, I believe, earlier than
Alexander[1]. The oldest Greek portrait-statue
known to me is that of Mausolus, now in the
British Museum; but the majority of the statues
of Greek philosophers and others are probably to
be referred to the Roman times, when the for-
mation of portrait-galleries became a favourite
pursuit. With the Greeks it was otherwise; the
ideal was ever uppermost in their mind: they
executed busts of Homer indeed and placed his
head on many of their coins; but of course these
were no more portraits than the statues of Jupiter and Pallas are portraits. With regard to the
relation of Greek archæology to the history of
Greece, both the monuments and the literature are

[1] I am aware that there are reasons for believing that a
Persian coin preserves a portrait of Artaxerxes Mnemon, who
reigned a little earlier.

abundant, and they mutually illustrate one another; and the same remark is more or less true for the histories of the nations afterwards to be mentioned, upon which I shall therefore not comment in this respect.

From Greece, who taught Rome most or all that she ever knew of the arts, we pass to the contemplation of the mistress of the world herself. She found indeed in her own vicinity an earlier civilisation, the Etruscan, whose archæological remains and history generally are amongst the most obscure and perplexing matters in all the world of fore-time. The sepulchral and other monuments of Etruria are often inscribed, but ro ingenuity has yet interpreted them. The words of the Etruscan and other Italian languages have been recently collected by Fabretti. There is some story about a learned antiquary after many years' research coming to the conclusion that two Etruscan words were equivalent to *vixit annos*, but which was *vixit*, and which *annos*, he was as yet uncertain. We have also Etruscan wall-paintings, and various miscellaneous antiquities in bronze, and among them the most salient peculiarity of Etruscan archæology not easily to be conjectured, its elegantly-formed bronze mirrors. These, which are incised with mythological subjects, and often inscribed, have attracted the especial attention of modern scholars and antiquaries, who have gazed upon them indeed almost as wistfully as the Tuscan ladies themselves.

But Greece had far more influence over Roman
life and art than Etruria.

> Græcia capta ferum victorem cepit, et artes
> Intulit agresti Latio.

Accordingly, Greek architecture (mostly of the
later Corinthian style, which was badly elabo-
rated into the Composite) was imported into Rome
itself, and continued to flourish in the Greek
provinces of the empire. Temples and theatres
continued much as before; but the triumphal arch
and column, the amphitheatre, the bath and the
basilica, are peculiarly Roman.

The genius of Rome however was essentially
military, and the stamp which she has left on the
world is military also.   Her camps, her walls,
and her roads, *strata viarum*, which, like arteries,
connected her towns one with another and with
the capital, are the real peculiarities of her archæology. The treatise on Roman roads, by Bergier,
occupies above 800 pages in the *Thesaurus* of
Grævius. Instead of bootlessly wandering over the
width of the world on these, let us rather walk
a little over those in our own country, and as we
travel survey the general character of the Roman
British remains, which may serve as a type of all.
In the early part of this lecture, I observed that
we, in common with the rest of Western Europe,
find in our islands weapons which belong to the
stone, bronze, and iron periods; and here also,
as in other places, the last-named period doubtless
connects itself with the Roman. But besides

these, we have other remains, many of which may
be referred to the Celtic population which Cæsar
had to encounter, when he invaded our shores.
These remains may in great part perhaps (for
I am compelled to speak hesitatingly on a subject which I have studied but little, and of which
no one, however learned, knows very much) be
anterior to Roman times. Of this kind are
the cromlechs at Dufferin in South Wales, in
Anglesey, and in Penzance, of which there are
models in the British Museum; of this kind also
are, most probably, the gigantic structures at
Stonehenge, about which so much has been
written and disputed. The British barrows of
various forms and other sepulchral remains may
also be referred, I should conceive, in part at least,
to the pre-Roman Celtic period. The earlier
mounds contain weapons and ornaments of stone,
bronze and ivory, and rude pottery; the later ones,
called Roman British barrows, appear mostly not
to contain stone implements, but various articles
of bronze and iron and pottery; also gold ornaments and amber and bead necklaces. Other sepulchral monuments consist merely of heaps of stones
covering the body which has been laid in the
earth. Many researches into this class of remains
have of late years been made, and by none perhaps more patiently and more successfully than by
the late Mr Bateman, in Derbyshire. The archæology of Wales has also been made the special object of study by a society formed for the purpose.

Some tribes of the ancient Britons were certainly acquainted with the art of die-sinking, and
a great many coins, principally gold, are extant,
some of which may probably be as early as the
second century before Christ. They are, to speak
generally, barbarous copies of the beautiful gold
staters of Philip of Macedon, which circulated
over the Greek world, and so might become
known to our forefathers by the route of Marseilles.

With these remarks I leave the Celtic remains in
Britain; all attempts to connect together the literary notices and the antiquities of the Celts and
Druids, so as to make out a history from them,
have been compared to attempts to " trace pictures
in the clouds[1]." Still we may say to the Celtic
archæologist,

Οαρσεῖν χρή, φίλε Βάττε, τάχ᾽ αὔριον ἔσσετ᾽ ἄμεινον.

One day matters may become clearer by the
help of an extended and scientific archæology.

But of the Romano-British remains it may be
necessary to say something. When we look at
the map in Petrie's *Monumenta Historica Britannica*, in which the Roman roads are laid down
by their actual remains, we see the principal
Roman towns and stations connected together by
straight lines, which are but little broken. So numerous are they that we might almost fancy that
we were looking at a map in an early edition of a

---

[1] *Pict. Hist. of England*, Vol. I. p. 59. London, 1837.

Railway Guide. In this county they abound and
have been very carefully traced, and both here
and in other counties are still used as actual roads.
In a few instances mile-stones have also been
found. In our own country, cut off, as Virgil
says, from the whole world, we do not expect the
splendid monuments of Roman greatness, yet
even here the temple, the amphitheatre and the
bath are not unknown; and in our little Pompeii at Wroxeter we have, if my memory deceive
me not, some vestiges of fresco-painting, an art
of which we have such beautiful Roman examples
elsewhere. But everywhere we stumble upon
camps and villas; everywhere

> The tesselated pavements shew
> Where Roman lamps were wont to glow.

And of these lamps themselves we have an infinite
number and variety, and on many of them representations of the games of the circus and of various other things, formed in relief; a remark which
may also be made of their fine and valuable red
Samian ware; fragments of which are commonly
met with, but the vases are rarely entire. Of
their other pottery, and of their glass and per-
sonal ornaments, and miscellaneous objects, I must
hardly say any thing; but only observe that the
Romans have left us a very interesting series of
coins relating to Britain; Claudius records in gold
the arch he raised in triumphant victory over us:
in the same way Hadrian, Antoninus Pius, Sep-
timius Severus, besides building their great walls

against us, have, as well as Caracalla and Geta,
struck many pieces in silver and copper to com-
memorate our tardy subjugation. The British
emperors or usurpers, Carausius and Allectus,
have also left us very ample series of coins, and
indeed it is by these, much more than by the
monuments of letters, that their histories are
known. In the fourth and fifth centuries the
monetary art declined greatly in the Western
Empire, and was on the whole at a very low ebb
in the Eastern or Byzantine Empire, and in the
middle ages, generally, throughout Europe.

At Constantinople a new school of Roman
art arose, which exercised a powerful influence on
medieval art in general. Soon after the foundation of Constantinople, Roman artists worked there
in several departments with a skill by no means
contemptible, though of a strangely conventional
and grotesque character; and from them, as it
would seem, the medieval artists of Central and
Western Europe caught the love of the same
crafts, and carried them to much higher excellence.
I would allude in the first place, as being among
the earliest, to ivory carvings, principally consular
diptychs. From the time of the emperors it was
the custom for consuls and other curule magis-
trates to make presents both to officials and their
friends of ivory diptychs, which folded together
like a pair of book-covers, on which sculptures
in low relief were carved, as a mode of announcing
their elevation. From the fourth and fifth centu-

ries down to the fourteenth we find them, some of
the earliest with classical subjects, as the triumph
of Bacchus, probably of the fourth century; but
mostly with Scriptural ones, or with representations of consuls. Some of these are enriched with
jewellery. The inscriptions accompanying them
are either in Greek or in Latin. In Germany they
occur in the Carlovingian period, though rarely,
and in France and Italy later still. Perhaps it
should be mentioned that the ivory episcopal chair
of St Maximian at Ravenna, a work of the sixth
century, is the finest example extant of this class
of antiques, and is doubly interesting as being one
of the very few extant specimens of furniture
during the first three centuries of the middle ages.
Various casts of medieval ivories, it may be
added, have been executed and circulated by the
Arundel Society.

Another art learnt from Rome in her decline,
or from Constantinople, is the illumination of
MSS., which the calligraphers of the middle ages
in all countries throughout Europe carried to a
very high perfection. Perhaps the earliest example to be named is the Greek MS. of Genesis
in the LXX, now preserved in the Imperial Library
at Vienna, probably of the fourth century. The
vellum is stained purple, and the MS. is decorated
with pictures executed in a quaint, but vigorous
style. In these, we find (as M. Labarte[1], a great

[1] *Histoire des Arts au moyen âge.* Album. Vol. II. pl.
lxxvii. Paris, 1864.

authority for medieval art, assures us) all the
characters of Roman art in its decline, such as it
was imported to Constantinople by the artists
whom Constantine called to his new capital; and
"they have served," as he adds, "for a point of
departure" in the examination which he has made
of the tendencies and destinies of Byzantine art.
Compare the Vatican MSS. of Terence and Virgil.
I cannot be expected to enter into details about
illuminations; they occur in MSS. of all sorts,
more or less, in Europe, down to the sixteenth
century, but especially in sacred books, such as
were used in Divine service. I need only call to
your remembrance the beautiful assemblage exhibited in the Fitzwilliam Museum and in the
University Library, to say nothing of the treasures possessed by our different colleges.

There are many other objects of medieval art
not unworthy of being enlarged upon, which I in-
tentionally pass over lightly, lest their multiplicity
should distract us; thus I will say little of its
pottery, its coins, or of its sculptures and bas-
reliefs in stone. With regard to the first of them,
M. Labarte observes: "It is not until the beginning of the fifteenth century that we find among
the European nations any pottery, but such as
has been designed for the commonest domestic
use, and none that art has been pleased to deco-
rate." These are objects which the middle ages
have in common with others; and they are objects
in which a comparison will not be favourable to

medieval art. Still, we must take care that a love of art does not blind us to the real value of such things; they are always interesting for the *history* of art, whatever their rudeness or whatever their ugliness; and, moreover, they are often, as the coins of various nations, of high historical interest. For example, on our own series of barbarous Saxon coins we have not only the successions of kings handed down to us, in the several kingdoms of the so-called Heptarchy and in the united kingdom, but also on the reverses of the same coins we have mention made of a very large number of cities and towns at which they were respectively struck. For example, to take Cambridge, we find that coins were struck here by King Edward the Martyr, Ethelred the Second, Canute, Harold the First, and Edward the Confessor; also after the Conquest by William the First and William the Second. We are thus furnished with very early notices, and so in some measure able to estimate the importance of the cities and towns of our island in medieval times; though great caution is necessary here in making deductions; for no coins appear to have been struck in Cambridge after the reign of William Rufus. And this seems at first sight so much the more surprising when we bear in mind that money was struck in some of our cities, as York, Durham, Canterbury, and Bristol, quite commonly, as late as the fifteenth and sixteenth centuries. But, in truth, from the twelfth century downwards, the

4

number of cities and towns in which lawful money
was struck became comparatively small.

But I must not wander too far into numismatics. The art of enamelling, peculiarly characteristic of the later periods of the middle ages,
is very fully treated of by M. Labarte, from whom
I derive the following facts. The most ancient
writer that mentions it is the elder Philostratus,
a Greek writer of the third century, who emi-
grated from Athens to Rome. In his *Icones*, or
Treatise on Images, the following passage occurs.
After speaking of harness enriched with gold,
precious stones, and various colours, he adds: "It
is said that the barbarians living near the ocean
pour colours upon heated brass, so that these
adhere and become like stone, and preserve the
design represented." It may, therefore, be considered as established that the art of enamelling
upon metals had no existence in either Greece or
Italy at the beginning of the third century; and,
moreover, that this art was practised at least as
early in the cities of Western Gaul. During the
invasions and wars which desolated Europe from
the fourth to the eleventh century almost all the
arts languished, and some may have been entirely
lost. Enamelling was all but lost; for between the
third and the eleventh centuries the only two
works which occur as landmarks are the ring of
King Ethelwulf in the British Museum, and the
ring of Albstan, probably the bishop of Sherburne,
who lived at the same time. These two little

pieces, however, only serve to establish the bare existence of enamelling in the West in the ninth century. But in this same century the art was in all its splendour at Constantinople, and we possess specimens of Byzantine workmanship of even an earlier date. I cannot enter into the various modes of enamelling, which are fully described by M. Labarte; but merely mention, without comment, a few of the principal specimens, independently of the Limoges manufacture, which constituted the chief glory of that city from the eleventh century to the end of the medieval period. "This became the focus whence emanated nearly all the beautiful specimens of enamelled copper, which are so much admired and so eagerly sought after for museums and collections." The principal earlier examples then are these; the crown and the sword of Charlemagne, of the ninth century, now in the Imperial Treasury at Vienna; the chalice of St Remigius, of the twelfth century, in the Imperial Library at Paris; the shrine of the Magi in Cologne, and the great shrine of Nôtre Dame at Aix-la-Chapelle, presented by the Emperor Frederick Barbarossa in the latter part of the same twelfth century. Also the full-length portrait (25 inches by 13) of Geoffrey Plantagenet, father of our Henry II., which formerly ornamented his tomb in the cathedral, but is now in the Museum at Le Mans. The British Museum likewise contains two or three fine examples; and among them an enamelled plate representing Henry of Blois,

4—2

Bishop of Winchester, and brother of King Stephen.

Very fine also are the extant products of the goldsmith's art in the middle ages; which date principally from the eleventh century, when the art received a new impulse in the West; those of earlier date, with very few exceptions, now cease to exist. They are principally chalices, reliquaries, censers, candlesticks, croziers and statuettes.

Nor can I pass over in absolute silence the armour of the middle ages. Until the middle of the ninth century it would appear to have resembled the Roman fashion, of which it is needless to say anything; but in Carlovingian times the hilts and scabbards of dress-swords were very highly decorated; and about this period, or rather later, the description of armour used by the ancients was exchanged for the hauberk or coat of mail, which was the most usual defensive armour during the period of the Crusades. The first authentic monument where this mail-armour is represented is on the Bayeux tapestry of Queen Matilda, representing the invasion of England by William Duke of Normandy in 1066; the most famous example of medieval tapestry in existence, though other specimens are to be seen at Berne, Nancy, La Chaise Dieu, and Coventry. The art of the *tapissier*, however, in the eleventh century, when the Bayeux tapestry was made, would appear to have been on the decline.

In the beginning of the fourteenth century plate-armour began to come into use; and by and by this was decorated with Damascene work, a style of art applied to the gate of a basilica in Rome, which was sent from Constantinople, as early as the eleventh century, but which did not become general in the West till the fifteenth. To this I may just add, that sepulchral brasses, on which figures in armour are often elaborately represented by incised lines, are a purely medieval invention of the thirteenth century. Sir Roger de Trumpington's brass at Trumpington is one of the very earliest examples. But time forbids me to say more of sepulchral brasses, a class of antiquities almost confined to our own country, of which we have some few specimens as late as the seventeenth century, or to do more than allude to the beautiful sepulchral monuments in stone of the medieval period, with which we are all more or less familiar.

The most remarkable art to which the middle age gave birth was oil-painting, the very queen of all the fine arts, though it was to the age of the Medici that its immense development was due. Previously painting had been subordinated to architecture; but now, while mosaics, frescoes, and painted glass remained still subservient to her, the art of painting occupies a distinct and prominent rank of its own. It used commonly to be said that the invention of painting on prepared panel was due to Margaritone of Arezzo, who died

about 1290, and in like manner that John van
Eyck invented oil-painting in 1410. Both these
errors have been propagated by the authority of
Vasari. But it is now well known, and has been
conclusively proved, both by M. Labarte and by
Sir C. Eastlake, that these modes of painting
are mentioned by authors who lived more than
a century before Margaritone, in particular by
the monk Theophilus, who in the twelfth century
composed a work entitled *Diversarum artium
schedula*. Paintings in oil either are or lately
were in existence anterior to John van Eyck;
for example one at Naples, executed by Filippo
Tesauro, and dated 1309. We must ascend to
much earlier times to discover the true origin
of portable paintings, and we shall find it in the
Byzantine Empire. The Greeks, about the time
that the controversy respecting images was rife,
multiplied little pictures of saints; these were
afterwards brought over in abundance by the
priests and monks who followed the crusades,
and from the study of them, schools of painting
in tempera arose in Italy, in the twelfth century,
at Pisa, Florence and other places. The Byzantine school, M. Labarte tells us, reigned para-
mount in Italy until the time of Giotto, *i. e.* the
beginning of the fourteenth century, and also in
the schools of Bohemia and Cologne, the most
ancient in northern Europe, until towards the end
of the fourteenth century. In this country we
have two very early paintings, one of the be-

ginning and the other of the end of the same
fourteenth century, in Westminster Abbey. The
former, probably a decoration of the high altar,
is on wood; it represents the Adoration of the
Magi and other Scriptural subjects, and is declared by Sir C. Eastlake to be worthy of a good
Italian artist of the fourteenth century, though
he thinks that it was executed in England.
The latter is the canopy of the tomb of Richard
II. and Anne his first wife, representing the
Saviour and the Virgin and other figures. The
action and expression are declared by Sir C.
Eastlake to indicate the hand of a skilful
painter. In 1396, £20 was paid by the sacrist
for the execution of the work. These remarks
must suffice for a notice of medieval painting;
the glorious period of its history belongs rather to
the Renaissance, or post-medieval age.

The only archæological monuments of great
importance which remain to be mentioned are
those of architecture, in connection with the ac-
cessories of mosaics, frescoes, and painted glass.
The two former descended from classical times,
the last is the creation of the middle age.
Mosaics having been originally used only in
pavements, at length were employed as embellish-
ments for the walls of basilicas, and, by a natural
transition, of churches. Constantine and his successors decorated many churches in this manner,
and in the East a ground of gold or silver was
introduced below the glass cubes of the mosaics,

and a lustre was by this means spread over the work
which in earlier times was altogether unknown.
Thus the tympanum above the principal door of
the narthex of the Church of St Sophia, built by the
Emperor Justinian at Constantinople, is adorn-
ed with a mosaic picture of the Saviour seated,
the cubes of the mosaics being of silvered glass;
it is accompanied by Greek texts. This and
other later mosaics are figured by M. Labarte,
in his last and most splendid work, entitled
Histoire des Arts au moyen âge; among the rest
a Transfiguration of the tenth century. The
Byzantine art, with its stiff conventionality, prevailed every where till Cimabue, G. Gaddi, and
Giotto imparted to its rudeness a grace and noble-
ness which marked a new era. In the vestibule
of St Peter is a noble mosaic, partly after the
design of Giotto, representing Christ walking on
the water, and the apostles in the ship. But
the very masters who raised the art to its perfection brought about its destruction. Painting,
restored by these same great men, was too powerful a rival; and after the sixteenth century, when
it still flourished in Venice under the encourage-
ment of Titian, we hear little more of mosaics on
any great scale.

Passing over frescoes, which were much en-
couraged by Charlemagne, and by various sove-
reigns and popes during the middle ages, because the ravages of time have either destroyed
them altogether or left them in a deplorable con-

dition, as for example in some parish-churches in England, I will make a few remarks on painted glass, so extensively used in the decoration of the later churches.

The art of painting glass was unknown to the ancients, and also to the early periods of the middle ages. "It is a fact," says M. Labarte, "acknowledged by all archæologists, that we do not now know any painted glass to which an earlier date than the eleventh century can be assigned with certainty." Two specimens, and no more, of this century, are figured by M. Lasteyrie. The painted windows of the twelfth and thirteenth centuries are nearly of the same character. They consist of little historical medallions, distributed over mosaic grounds composed of coloured (not painted) glass, borrowed from preceding centuries. Fine examples from the church of St Denys and La Sainte Chapelle at Paris, of the twelfth and thirteenth centuries, are figured by M. Lasteyrie, and also by M. Labarte, who has many beautiful remarks on their harmony with the buildings to which they belong, on the elegance of their form, the richness of their details, and the brilliancy of their colours. In the fourteenth century, when examples become common, the glass-painters copied nature with more fidelity, and exchanged the violet-tinted masses, by which the flesh-tints had been rendered, for a reddish gray colour, painted upon white glass, which approached more nearly to nature. Large single figures now often occupy

an entire window. The improvement in drawing
and colouring is a compensation for the more
striking effects of the brilliant yet mysterious
examples of the preceding centuries; and the end
of the fourteenth century is one of the finest
epochs in the history of painted glass. Painting
on glass followed the progress of painting in oils
in the age which followed; and artists more
and more aimed at producing individual works;
and in the latter half of the fifteenth century
buildings and landscapes in perspective were first
introduced. The decorations which surround the
figures being borrowed from the architecture of
the time have often a very beautiful effect. But
the large introduction of *grisailles* deprives the
windows of this period of the transparent brilliancy
of the coloured mosaics of the earlier glass-painting. In the sixteenth century, however, glass was
nothing more than the material subservient to
the glass-painter, like canvas to the oil-painter.
Small pictures very highly finished were executed
after the designs of Michael Angelo, Raffaelle,
and the other great painters of the Renaissance.
"But," as M. Labarte truly says, "the era of
glass-painting was at an end. From the moment
that it was attempted to transform an art of
purely monumental decoration into an art of expression, its intention was perverted, and this
led of necessity to its ruin. The resources of
glass-painting were more limited than those of
oil, with which it was unable to compete. From

the end of the sixteenth century the art was in
its decline, and towards the middle of the seventeenth was" almost "entirely given up." Our own
age has seen its revival, and though the success
has been indeed great, we may hope that the
zenith has not yet been reached. "It is," says
Mr Winston, "a distinct and complete branch of
art, which, like many other medieval inventions,
is of universal applicability, and susceptible of
great improvement." I have been a little more
diffuse on glass-painting than on some other
subjects, as it is a purely medieval art, and one
which has now acquired a living interest. Various
examples of the different styles will easily suggest themselves to many, or, if not, they may be
studied in the splendid work of M. Lasteyrie,
entitled *Histoire de la Peinture sur Verre d'après
ses monuments en France,* and on a smaller scale
in Mr Winston's valuable *Hints on Glass-painting.*

With regard to the architectural monuments of
the medieval world, I may, in addressing such
an audience, consider them to be sufficiently well
known for my present purpose, which is to give
an indication, and little more, of the archæological
remains which have come down to our own days.
Medieval architecture is in itself a boundless
subject; and as I have not specially studied it,
I could not, if I would, successfully attempt an
epitome of its various forms of Byzantine, Sara-
cenic, Romanesque, Lombardic, and of infinitely

diversified Gothic. For a succinct yet comprehensive view of all these and more, I must refer
you to Mr Fergusson's *Handbook of Architecture*.
Yet when we let our imagination idly roam over
Europe, and the adjoining regions of Asia and
Africa, what a host of architectural objects flits
before it in endless successions of variety and
beauty! Think of Justinian's Church of St Sophia, which he boasted had vanquished Solomon's
temple, and again of St Mark's at Venice, as
Byzantine examples. Think next of the mosque
of the Sultan Hassan, and of the tombs of the
Memlooks mingled with lovely minarets and
domes at Cairo; of the Dome of the Rock at
Jerusalem; of the Alhambra in Spain, with all
the witchery of its gold and azure decorations.
Float, if you will, along the banks of the Rhine
or the Danube (as many of us have actually done),
and conjure up the majestic cathedrals, the spa-
cious monasteries and the ruined castles, telling
of other days, with which they are fringed. Let
the bare mention of the names of Milan, Venice,
Rome; again of Paris, Rheims, Chartres, Amiens,
Troyes, Rouen, Avignon; and in fine those of
Antwerp, Louvain, and Brussels, suggest their
own stories. Yet the magnificent structures,
secular and ecclesiastical, which I have either
named or hinted at, need not make us ashamed
of our own country. We are surrounded on all
sides by an archæology which is emphatically an
archæology of progress, and we may justly be

proud of it as Englishmen. In this University and its immediate neighbourhood we have fine specimens of Saxon, Norman, Early English, Decorated, and Perpendicular styles of Gothic architecture; and as regards the last of them, one of the most splendid examples in the world. In the opinion of competent judges the English cathedrals, while surpassed in size by many on the Continent, are in excellence of art superior to those of France or of any country in Europe. "Nothing can exceed the beauty of the crosses which Edward I. erected on the spots where the body of Queen Eleanor rested on its way to London." Some of these, Waltham for example, are quite equal to anything of their class found on the Continent. "The vault of Westminster Abbey" (says Mr Fergusson, on whose authority I make almost every statement relating to mediæval architecture) "is richer and more beautiful in form than any ever constructed in France;" the triforium is as beautiful as any in existence; and its appropriateness of detail and sobriety of design render it one of the most beautiful Gothic edifices in Europe.

I thus conclude my sketch, such as it is, of the archæology of the world. Its aim has been to bring under review the rude implements and weapons of primeval man; the colossal structures of civilised man in Egypt and India; the strangely-compounded palace-sculptures of Assyria and Babylonia; the exquisitely ornamented columns of

Persian halls; the massive architecture of Phœnicia; the Gothic-like rock-tombs of Lycia; the
lovely temples, and incomparable works of art of
every kind, great and small, of Greece; the military impress of Roman conquest; the medieval
works of art in ivory, in enamel, in glass-painting,
as well as its glorious architectural remains, connecting the middle ages with our own times. It
has been drawn, as I observed at the outset, under
very adverse circumstances, and must on that
account venture to sue for much indulgence. It
is open, no doubt, to many criticisms: I expect
to be charged with grievous sins of omission,
and perhaps of commission also: nor do I suppose that I could entirely vindicate myself from
such charges. Worse than all perhaps, I have
exposed myself to the unanswerable sarcasm that
I have talked about many subjects of which
I know but little. If, however, I have been able
to compile from trustworthy sources or manuals
so much respecting those particular branches of
archæology which I have not studied, as to bring
before you their salient features in an intelli-
gible manner, that is enough for my purpose. I
want no more, and I pretend to no more; and I
am conscious enough that even this purpose has
been but feebly accomplished. Tediousness, indeed, in dealing with numerous details could hardly
be altogether avoided; but this is so much lighter
a fault than an indulgence in mere platitudes,
running smoothly and amusingly, but emptily

withal, that I shall hear your verdict of *guilty* with composure.

It now only remains that I should very briefly point out what qualifications are necessary for an archæologist, and also the pleasure and advantage which result from his pursuits.

With regard to the first of these matters, the qualifications necessary for an archæologist, they are to some considerable extent the same as are necessary for a naturalist.

Like the naturalist, the antiquary must in the first place bring together a large number of facts and objects. This is, no doubt, a matter of great labour, but believe me, *'labor ipsa voluptas.'* The labour is its own ample reward. The hunting out, the securing, and the amassing facts and objects of antiquity, or of natural history, are the field-sports of the learned or scientific Nimrod. In a certain sense every archæologist *must* be a collector; he must be mentally in possession of a mass of facts and objects, brought together either by himself or by others. It is not absolutely necessary that he should be a collector, in the sense of being owner of a collection of his objects of study; in some departments indeed of archæology to amass the objects themselves is impossible: who, for instance, can collect Roman roads or Gothic cathedrals? models, plans, and drawings, are the only substitutes possible. But, with the facts relating to his favourite objects, and also as much as possible with the objects themselves, he must be familiar.

Yet this familiarity will not be enough to make him an archæologist. Such knowledge may be possessed, and very often is possessed, by a mere dealer in antiquities. The true antiquary must not only be well acquainted with his facts, but he must also, when there are sufficient data, proceed to reason upon them. He puts them together, and considers what story they have to render up. We saw a beautiful illustration of this in the joint labours of the Scandinavian antiquaries and naturalists. The order and sequence of the stone, bronze, and iron ages, were distinctly made out; and even their chronology may one day be discovered. The antiquary is enabled to form some judgment of the civilisation, the arts, and the religion of the nations whose remains he studies. Very often, as in the Roman series of coins, he makes out political events in their history, and assigns their dates. He determines the place of things in the historical series, much as the naturalist does in the natural series.

Like the naturalist also he must be a man of learning, i. e. he must be acquainted with what has been written by his fellow-labourers in the same branch of study. Few know, prior to experience, what a serious business this is. The bibliography of every department of archæology, as well as of natural history, is now becoming immense.

But besides a knowledge of facts, and objects, and books, there are one or two other qualifications necessary for many departments of archæology, the want of which has been very prejudicial to

some distinguished writers. Exact scholarship is
one of these qualifications. I do not merely
mean that if a man be engaged in Greek archæology, he must be aware of the passages of Greek
authors, in which the vases or the coins he is talk-
ing about are alluded to, though he must certainly
be acquainted with these, and possess sufficient
scholarship to construe them correctly; but he
must also be able to interpret his written archæological monuments, such as his inscriptions and the
legends of his coins. This is oftentimes no easy
matter, and it requires a knowledge of strange words
and dialects. Moreover, if an inscription or a le-
gend be mutilated (and this is very frequently the
case), unless the archæologist has an accurate knowledge of the language in which it is written, whatever that may be, Greek, Latin, Norman-French,
or any other, what hope is there that he will
ordinarily be able to restore it, and having so
done interpret it with security or satisfaction?
As one illustration of many, I will cite Prof.
Ramsay's remark on Nibby's dissertation *Delle vie
degli Antichi:* "In the first part of this article (on
Roman roads) his essay has been closely followed.
*Considerable caution, however, is necessary in
using the works of this author,* who, although a
profound local antiquary is by no means an accurate scholar[1]." Mr Bunbury, while pointing out
the advantages which scholars would derive from

---

[1] See Smith's *Dict. Gr. and Rom. Antiq.* a. v. Via.

5

some acquaintance with archæology, points out
by implication the advantage which archæologists
would derive from scholarship. "In this country,"
says he, "the study of archæology is but too much
neglected; it forms no part of the ordinary training of our classical scholars at the Universities,
and is rarely taken up by them in after life. It is
generally considered as the exclusive province of
the professed antiquarian, who has seldom under-
gone that early training in accurate scholarship,
which is regarded, and we think with perfect jus-
tice, by the student from Oxford or Cambridge,
as the indispensable foundation of sound classical
knowledge[1]." I think he is a little over-severe on
us; living men like Mr C. T. Newton, Mr Waddington, Mr Vaux, Mr C. W. King, Mr C. K.
Watson, and, last, but not least, like himself, to
whom others might be added, prove that his asser-
tions must be taken *cum grano;* even if it be true
that this country has produced no work connected
with ancient art which can be compared with
the writings of Gerhard, or Welcker; of Thiersch,
or Karl Otfried Müller[2].

---

[1] *Edinburgh Review,* u. s.

[2] I feel a little inclined to dispute this: Stuart, one of
the authors of the *Antiquities of Athens,* which have been
continued by other very able hands, and have also been translated into German, may, perhaps, take rank with the authors
named in the text. K. O. Müller himself calls Millingen's
Ancient Unedited Monuments (London, 1822) "a model of a
work;" and though without doubt Millingen is inferior to
Müller in scholarship and in acquaintance with books, he is

Another thing very desirable for the successful prosecution of some branches of archæology
is an appreciation of art. Without it we cannot
judge of the value of many antiques, or enter into
their spirit or feeling; we neither discern their
excellencies nor their deficiencies. Mr King, who
has made the province of ancient gems peculiarly
his own, justly calls them "little monuments of
perfect taste,...only to be appreciated by the edu-
cated and practised eye[1]." Moreover, this is the
very knowledge often so requisite for distinguish-
ing genuine antiquities from modern counterfeits.
The modern forgers, who fabricate Greek coins from
false dies, do not often reach the freedom and
beauty of the originals; though it must be confessed that some of them, as Becker, have carried
their execrable art to a very high perfection. It
is but rarely that these men meet with the
punishment they deserve; yet it is satisfactory to
know that Charles Patin, great scholar and great
antiquary as he was, was banished by Lewis XIV.
from his court for ever, for selling him a false
coin of Otho; and that a manufacturer of antiques
in the East, near Bagdad I believe, lately received
by order of the Turkish governor a sound basti-
nado on the soles of his feet for reproducing the
idols of misbelievers of old time.

probably at least his equal as a practical archæologist. Colonel
Leake's *Numismata Hellenica* (London, 1856) may also be cited
as an admirable combination of learning with practical archæology.

[1] *Antique Gems*, Introd. p. xxiii. London, 1860.

A knowledge of natural history in fine is occasionally very useful to an antiquary. I will give two instances, not at all generally known, one taken from zoology, one from botany. On the reverse of the splendid Greek coins of Agrigentum a crab is commonly represented. To an ignorant eye the crab looks much like the crab in our shops here in Cambridge; the zoologist recognises in it the fresh-water crab of the regions of the Mediterranean; the numismatist, profiting by this knowledge, sees at once that the type of the coin symbolizes not the harbour of Agrigentum, as he had supposed, but its river. Again, on the reverse of the beautiful Greek coins of Rhodes occurs a flower, about which numismatists have disputed since the time of Spanheim, whether it was the flower of the rose or of the pomegranate. Even Col. Leake has here taken the wrong side, and decided in favour of the pomegranate; the divided calyx at once shews every botanist that the representation is intended for the rose, conventional as that representation may be, from which flower the island derives its name.

These are, I think, the principal qualifications which are necessary or desirable for the archæologist. It only remains that I should point out briefly some of the pleasures and advantages that result from his pursuits. For I shall not so insult any one of you, who are here present, as to suppose that this question is lurking secretly in your mind, "Is there any good in archæology at all?

To what practical end do your researches tend?"
My learned predecessor well says that "this question is sometimes put to the lover of science or
letters by those from whom nature has withheld
the faculty of deriving pleasure from the exercise
of the intellect, and he feels for the moment
degraded to the level of such." It is not so clear
however that the fault must be put to the account
of nature. Rather, we may say,

Homine *imperito* nunquam quidquam injustius,
Qui nisi quod ipse facit, nihil rectum putat.

" No one," says a Swedish scholar of the seventeenth century, "blames the study of antiquity
without evidencing his own ignorance; as they
that esteem it do credit to their own judgment;
so that to sum up its advantages we may assert,
there is nothing useful in literature, if the knowledge of antiquity be judged unprofitable[1]." It
is doubtless one of the many charms of archæology
that it illustrates and is illustrated by literature;
indeed, some knowledge of antiquity is little less
than necessary for every man of letters. Unless
we have some knowledge of the objects whose
names occur in ancient literature, we lose half the
pleasure of reading it. In reading the New Testament, I can certainly say for myself, that I derive
more pleasure from the narrative of the woman
who poured the contents of the alabaster box

[1] Figrelius, quoted in the *Museum of Classical Antiquities*,
Vol. L p. 4.

over the head of Jesus, now that I know what an
alabastron is, and how its contents would be ex-
tracted; and in the same way I appreciate the
remark made by the silversmith in the Acts, that
all Asia and the world worshipped the Ephesian
Diana, now that I know her image to be stamped
not on the coins of Ephesus only, but on many
other cities throughout Asia also.   Here, I think,
we have pleasure and profit combined in one.
Instances are abundant where monuments illustrate
profane authors. The reader of Aristophanes will
be pleased to recognise among the earliest figures
on vases that of the ἰπποαλεκτρύων, the cock-horse,
or horse-cock, which cost Bacchus a sleepless night
to conceive what manner of fowl it might be.
"The Homeric scholar again," it has been said,
"must contemplate with interest the ancient pictures of Trojan scenes on the vases, and can hardly
fail to derive some assistance in picturing them to
his own imagination, by seeing how they were
reproduced in that of the Greeks themselves in
the days of Æschylus and Pindar[1]."

Further, not only is ancient literature, but also
modern art, aided by archæology. It is well
known how, in the early part of the thirteenth
century, Niccola Pisano was so attracted by a bas-
relief of Meleager, which had been lying in Pisa
for ages unheeded, "that it became the basis of
his studies and the germ of true taste in Italy."

[1] *Edinburgh Review*, u. s.

In the Academy of St Luke at Rome, and in the schools established shortly afterwards at Florence by Lorenzo de' Medici, the professors were required to point out to the students the beauty and excellence of the works of ancient art, before they were allowed to exercise their own skill and imagination. Under the fostering patronage of this illustrious man and of his not less illustrious son a galaxy of great artists lighted up all Europe with their splendour. Leon Batista Alberti, one of the greatest men of his age, and especially great in architecture, was most influential in bringing back his countrymen to the study of the monuments of antiquity. He travelled to explore such as were then known, and tells us that he shed tears on beholding the state of desolation in which many of them lay. The prince of painters, Raffaelle,

> timuit quo sospite vinci
> Rerum magna parens et moriente mori,

and the prince of sculptors, Michael Angelo, both drew their inspiration from the contemplation of the art-works of antiquity. The former was led to improve the art of painting by the frescoes of the baths of Titus, the latter by the sight of a mere torso imbibed the principles of proportion and effect which were so admirably developed in that fragment[1]. And not only the arts of sculpture

---

[1] For this and the preceding facts see the *Museum of Classical Antiquities*, Vol. I. pp. 13—15. The frescoes of the baths

and painting, but those which enter into our daily
life, are furthered by the wise consideration of the
past. Who can have witnessed the noble exhibi-
tions in Hyde Park or at Kensington without
feeling how much the objects displayed were in-
debted to Hellenic art? In reference to the former
of these Mr Wornum says : "Repudiate the idea
of copying as we will, all our vagaries end in a
recurrence to Greek shapes; all the most beautiful
forms in the Exhibition, (whether in silver, in
bronze, in earthenware, or in glass,) are Greek
shapes; it is true often disfigured by the accessory
decorations of the modern styles, but still Greek in
their essential form[1]."

And yet I must, in concluding this Introductory Lecture, most strongly recommend to you the
study of archæology, not only for its illustration of
ancient literature, not only for its furtherance of
modern art, but also, and even principally, for its
own sake. "Hæc studia adolescentiam alunt, se-
nectutem oblectant, secundas res ornant, adversis
perfugium ac solatium præbent; delectant domi,
non impediunt foris, pernoctant nobiscum, pero-
grinantur, rusticantur[2]." Every one who follows a
pursuit in addition to the routine duties of life has,

---

of Titus have subsequently lost their brilliancy. See Quatre-
mère de Quincy's *Life of Raphael*, p. 263. Hazlitt's Trans-
lation. (Bogue's European Library).

[1] *The Exhibition as a Lesson in Taste*, p. xvii.* * *
(Printed at the end of the *Art-Journal Illustrated Catalogue*,
1851).

[2] Cicero *pro Archia poeta*, c. vii.

by so doing, a happiness and an advantage of which others know little. The more elevated the pursuit, the more exquisite the happiness and the more solid the advantage. Now if

The proper study of mankind is man,

then most assuredly archæology is one of the most proper pursuits which man can follow. For she is the interpreter of the remains which man in former ages has left behind him. By her we read his history, his arts, his civilisation; by her magical charms the past rises up again and becomes a present; the tide of time flows back with us in imagination; the power of association transports us from place to place, from age to age, suddenly and in a moment. Again the glories of the nations of the old world shine forth;

Again their godlike heroes rise to view,
And all their faded garlands bloom anew.

To adopt and adapt the words of one who is both a learned archæologist and a learned astronomer of this University, I feel that I may, under any and all circumstances, impress upon your minds the utility and pleasure of "every species and every degree of archæological enquiry." For "history must be looked upon as the great instructive school in the philosophical regulation of human conduct," as well as the teacher "of moral precepts" for all ages to come; and no "better aid can be appealed to for" the discovery, for "the confirmation, and for the demonstration of the facts of history, than the energetic pursuit of archæology"[1].

[1] See an address delivered at an Archæological meeting at Leicester, by John Lee, Esq., LL.D. (*Journal of Archæol. Association* for 1863, p. 37).

NOTES.

Pp. 15—20. Nearly everything contained in the text relating to prehistoric Europe will be found in the *Revue Archéologique* for 1864, and in Sir C. Lyell's *Antiquity of Man*, London, 1863; see also for Thetford, *Antiq. Commun.* Vol. I. pp. 339—341, (Cambr. Antiq. Soc. 1859); but the following recent works (as I learn from Mr Bonney, who is very familiar with this class of antiquities) will also be found useful to the student:

Prehistoric Times. By John Lubbock, F.R.S. London, 1865. 8vo.

The Primeval Antiquities of Denmark. By Prof. Worsäe. London, 1849. 8vo. (Engl. Transl.)

Les Habitations Lacustres. Par F. Troyon. Lausanne, 1860.

Les Constructions Lacustres du Lac de Neufchâtel. Par E. Desor. Neufchâtel, 1864.

Antiquités Celtiques et Antédiluviennes. Par Boucher de Perthes. Paris, 1847.

Die Pfahlbauten. Von Dr Ferd. Keller. Ber. I—V. (*Mittheilungen der Antiquarischen Gesellschaft in Zurich*). 1854, sqq. 4to.

Die Pfahlbauten in den Schweizer-Seen. Von I. Staub. Zurich, 1864. 8vo.

Besides these there are several valuable papers in the *Transactions of the Royal, Geological, and Antiquarian Societies* (by Messrs John Evans, Prestwich, and others), the *Natural History Review*, and other Periodicals.

p. 26. For the literature relating to ancient Egypt see Mr R. S. Poole's article on Egypt, in Smith's *Dictionary of the Bible*, Vol. I. p. 512.

pp. 29—31. Besides the works of Robinson, De Saulcy, Lewin, Thrupp, and others, the following books may be mentioned as more especially devoted to the archæology of Jerusalem:

The Holy City. By George Williams, D.D. (Second edition, including an architectural History of the Church of the Holy Sepulchre by the Rev. Robert Willis, M.A., F.R.S. 1849.)

Jerusalem Explored. By Ermete Pierotti. Translated by T. G. Bonney, M.A. 1864.

Le Temple de Jérusalem. Par le Comte Melchior de Vogüé, 1865. The Count considers none of the present remains of the Temple to be earlier than the time of Herod.

To these I should add Mr Williams' and Mr Bonney's tracts, directed against the views of Mr Fergusson, in justification of those of Dr Pierotti.

p. 31, l. 20. From some remarks made to me by my learned friend, Count de Vogüé, I fear that this is not so certain a characteristic of Phœnician architecture as has been commonly supposed. He assigns some of the bevelled stones which occur in Phœnicia to the age of the Crusades.

p. 31, last line. For the very remarkable Phœnician sarcophagus discovered in 1855, and for various references to authorities on Phœnician antiquities, see Smith's *Dict. of the Bible*, Vol. II. p. 868, and Vol. III. p. 1830.

p. 36. As a general work on Greek and Roman Coins Eckhel's *Doctrina Nummorum Veterum* (Vindobonæ, 1792—1828, with Steinbuchel's *Addenda*, 8 Vols. 4to.) still remains the standard, though now getting a little out of date.

The same remark must be made of Mionnet's great work, *Description de Médailles Antiques, Grecques et Romaines*, Paris, 1806—1813 (7 Vols.), with a supplement of 9 Vols. Paris, 1818—1837, giving a very useful *Bibliothèque Numismatique* at the end; to which must be added his *Poids des Médailles Grecques*, Paris, 1839. These seventeen volumes comprise the

Greek coins: the other part of his work, *De la Rareté et des Prix des Médailles Romaines*, Paris, 1827, in two volumes, is now superseded.

Since Mionnet's time certain departments of Greek and other ancient numismatics have been much more fully worked out, especially by the following authors:

De Luynes (coins of Satraps; also of Cyprus); L. Müller (coins of Philip and Alexander; of Lysimachus; also of Ancient Africa); Pinder (Cistophori); Beulé (Athenian coins); Lindsay (Parthian coins); Longpérier, and more recently Mordtmann (coins of the Sassanidæ); Carelli's plates described by Cavedoni (coins of Magna Græcia, &c.); other works of Cavedoni (Various coins); Friedländer (Oscan coins); Sambon (coins of South Italy); De Saulcy, Levy, Madden (Jewish coins); V. Langlois (Armenian, also early Arabian coins); J. L. Warren (Greek Federal coins; also more recently, copper coins of Achæan League); R. S. Poole (coins of the Ptolemies); Waddington (Unedited coins of Asia Minor).

For Roman and Byzantine coins (including Æs grave and Contorniates) see the works of Marchi and Tessieri, Cohen, Sabatier, and De Saulcy.

Others, as Prokesch-Osten, Leake, Smyth, Hobler, and Fox, have published their collections or the unedited coins of them; and all the numismatic periodicals contain various previously unedited Greek and Roman and other ancient coins.

p. 40. Fabretti's work is entitled, *Glossarium Italicum in quo omnia vocabula continentur ex Umbricis, Sabinis, Oscis, Volscis, Etruscis, cæterisque monumentis collecta, et cum interpretationibus variorum explicantur* (Turin, 1858—1864). Many figures of the antiquities, on which the words occur, are given in their places.

p. 43. Cromlechs in some, if not in all cases, appear to be the skeletons of barrows.

p. 44. The following works will be found useful for the student of early British antiquities:

Pictorial History of England, Vol. I. Lond. 1838.

Archæological Index to remains of Antiquity of the Celtic, Romano-British, and Anglo-Saxon periods. By J. Y. Akerman, F.S.A. London, 1847 (with a classified index of the Papers in the *Archæologia*, Vols. I—XXXI).

Ten years' diggings in Celtic and Saxon Grave Hills in the Counties of Derby, Stafford, and York, from 1848—1858. By Thomas Bateman. London, 1861. A most useful work, which will indicate the existence of many others. In connection with this see Dr Thurnam's paper on British and Gaulish skulls in *Memoirs of Anthropological Soc.* Vol. ı. p. 120.

The Land's End District, its Antiquities, Natural History, &c. By Richard Edmonds. London, 1862.

Catalogue of the Antiquities of Stone, Earthen, and Vegetable Materials, in the Museum of the Royal Irish Academy. By W. R. Wilde, M.R.I.A. Dublin, 1857.

The Coins of the Ancient Britons. By John Evans, F.S.A. The plates by F. W. Fairholt, F.S.A. London, 1864. By far the best and most complete work hitherto published on the subject.

Also, the *Transactions* of various learned Societies in Great Britain and Ireland, among which the *Archæologia Cambrensis* is deserving of special mention.

- For the Romano-British Antiquities may be added Horsley's *Britannia Romana*, 1732; Roy's *Military Antiquities of the Romans in Britain*, 1793; Lysons' *Relliquiæ Britannico-Romanæ*. London, 1813, 4 Vols. fol.

Monographs on York, by Mr Wellbeloved; on Richborough and other towns, by Mr C. R. Smith; on Aldborough, by Mr H. E. Smith; on Wroxeter, by Mr Wright; on Caerleon, by Mr Lee; on Cirencester, by Messrs Buckman and Newmarch; on Hadrian's wall, by Dr Bruce; on various excavations in Cambridgeshire, by the Hon. R. C. Nevilla.

p. 45. For the Roman Roads, &c. in Cambridgeshire, see Prof. Charles C. Babington's *Ancient Cambridgeshire*, Cambr. 1853 (Cambr. Ant. Soc.).

— No doubt need have been expressed about Wroxeter, which should hardly have been called 'our little Pompeii'; the area of Wroxeter being greater, however less considerable the remains. See Wright's *Guide to Uriconium*, p. 88. Shrewsbury, 1860. For various examples of Roman wall-painting in Britain see *Reliq. Isur.* by H. E. Smith, p. 18, 1852.

p. 46. For Romano-British coins see

Coins of the Romans relating to Britain, described and illustrated. By J. Y. Akerman, F.S.A. London, 1844.

Petrie's *Monumenta Historica Britannica*, Pl. i—xvii. London, 1848 (for beautiful figures).

Others, published by Mr C. R. Smith in his valuable *Collectanea Antiqua;* also by Mr Hobler, in his *Records of Roman History, exhibited on Coins.* London, 1860. Others in the *Numismatic Chronicle,* in the *Transactions of the Cambridge Antiquarian Society,* and perhaps elsewhere.

For medieval and modern numismatics in general we may soon, I trust, have a valuable manual (the MS. of which I have seen) from the pen of my learned friend, the Rev. W. G. Searle. He has favoured me with the following notes:

On medieval and modern coins generally we have

Appel, *Repertorium zur Münzkunde des Mittelalters und der neuern Zeit,* 6 Vols. 8vo. Pesth, 1820—1829.

Barthélemy, *Manuel de Numismatique du moyen âge et moderne.* Paris, 1851. 12mo.

The bibliography up to 1840 we get in

Lipsius, *Biblioth. Numaria,* Leipz. 1801 (2 Vols.) 8vo., and in

Leitzmann, *Verzeichniss aller seit 1800 erschienenen Numism. Werke,* Weissensee, 1841, 8vo.

On medieval coins, their types and geography, we have

J. Lelewel, *La Numismatique du Moyen-âge, considérée sous le rapport du type.* Paris, 1835, 2 vols. 8vo. Atlas 4to.

Then there are the great Numismatic Periodicals:

Revue Numism. 8vo. Paris, 1836.

Revue de la Num. Belge, 8vo. Brussels, 1841.

Leitzmann, *Numismatische Zeitung,* 4to. Weissensee, 1834.

On Bracteates:

Mader, *Versuch über die Bracteaten.* Prague, 1797, 4to.

And the great Coin Catalogues of

Welzl v. Wellenheim. 3 vols. 8vo. Vienna, 1844 ff. (c. 40,000 coins).

v. Reichel at St Petersburgh, in at least 9 parts.

On current coins we have

Lud. Fort, *Neueste Münzkunde,* engravings and descr. 8vo. Leipzig, 1851 ff.

p. 45. For almost everything relating to ivories and for a great deal on the subjects which follow, see *Handbook of the Arts of the Middle Ages and Renaissance,* Translated from the French of M. Jules Labarte, with notes, and copiously illus-

trated, London, 1855, which will lead the student to the great
authorities for medieval art, as Du Sommerard, &c. I have
also examined and freely used *Histoire des Arts industriels
au moyen âge et à l'époque de la Renaissance*, Par Jules Labarte.
Paris, 1864, 8vo. 2 volumes; accompanied by an album in
quarto with descriptions of the plates, also in two volumes.

p. 47. For examples of medieval calligraphy and illuminations
see Mr Westwood's *Palæographia Sacra Pictoria*, (Lond. 1845),
and his *Illuminated Illustrations of the Bible*, (London, 1840).

p. 48. A good deal of information about Celtic, Romano-
British, and medieval pottery will be found in Mr Jewitt's *Life
of Wedgwood*, London. 1865. For ancient pottery in general
(excluding however the medieval) see Dr Birch's *Ancient Pottery and Porcelain*, London, 1858, which will conduct the
student to the most authentic sources of information. In connection with this should be studied Mr Bunbury's article in
the *Edinburgh Review* for 1858, to which Mr Oldfield's paper
on Sir W. Temple's vases in the *Transactions of the Royal Soc.
of Lit.* Vol. vi. pp. 130—149 (1859), may be added.

— For medieval sculpture see Flaxman's *Lectures*.
The 'horrible and burlesque' style of the earlier ages was
discarded in the thirteenth century, when the art revived in
Italy. Italian artists executed various sepulchral statues in
this country, which possess considerable merit, as do others by
native artists, but the great beauty of our sepulchral monuments consists in their architectural decorations.

p. 49. For the coinage of the British Islands see the
works of Ruding, Hawkins, and Lindsay, also for the Saxon
coins found in great numbers in Scandinavia, Hildebrand
and Schröder. Humphreys' popular work on the coinage of
the British Empire, so far as the plates are concerned, is useful, but the author is deficient in scholarship.

p. 52. For the statements here made on oil-painting see
Bryan's *Dict. of Painters and Engravers*, by Stanley, (London,
1849), under Van Eyck, and Sir C. L. Eastlake's *Materials for
a History of Oil-painting.* (London 1847.)

p. 53. For medieval brasses, see

Bowtell, *Monumental Brasses and Slabs.* London, 1847, 8vo.

——— *Monumental Brasses of England, a Series of engravings in wood.* London, 1849.

Haines, *Manual of Monumental Brasses.* 2 parts. London, 1861, 8vo. This contains also a list of all the brasses known to him as existing in the British Isles. Mr Way has given an account of foreign sepulchral brasses in *Archæol. Journ.*, Vol. VII.

p. 56. Several English frescoes are described and figured in the *Journal of the Archæological Association*, passim.

p. 62, l. 13. The omission of ancient costume has been pointed out to me. The *actually existing* specimens however are mostly very late; with the exception of a few articles of dress found in Danish sepulchres of the bronze period, or in Irish peat bogs of uncertain date, the episcopal vestments of Becket now preserved at Sens are the earliest which occur to my recollection; and there are few articles of dress, I believe, so early as these. However both ancient and medieval costume is well known from the *representations* on monuments of various kinds. See *inter alia* Hope's *Costume of the Ancients;* Becker's *Gallus* and *Charicles;* Strutt's *Dress of the English People,* edited by Planché, (Lond. 1842); Shaw's *Dresses and Decorations of the Middle Ages.*

p. 67. The statement about Patin is made on the authority of a note in Warton's edition of Pope's Works, Vol. III. p. 306. (London 1797.)

p. 68. The remark about the crab was made to me by the late Mr Burgon, and I do not know whether it has ever been printed; its truth seems pretty certain. For the Rhodian symbol see my paper in the *Numismatic Chronicle* for 1864, pp. 1—6.

PREPARING FOR PUBLICATION,

An Introduction to the Study of Greek Fictile Vases; their Classification, Subjects, and Nomenclature. Being the substance of the Disney Professor's Lectures for 1865, and of those which he purposes to deliver in 1866.